551·6.

Items should be returned on or before the last date shown below. Items not already requested by other borrowers may be renewed in person, in writing or by telephone. To renew, please quote the number on the barcode label. To renew online a PIN is required. This can be requested at your local library.
Renew online @ **www.dublincitypubliclibraries.ie**
Fines charged for overdue items will include postage incurred in recovery. Damage to or loss of items will be charged to the borrower.

Leabharlanna Poiblí Chathair Bhaile Átha Cliatl
Dublin City Public Libraries

Baile Átha Cliath
Dublin City

Date Due	Date Due	Date Due
2 5 OCT 2019		

VERY SHORT INTRODUCTIONS are for anyone wanting a stimulating and accessible way into a new subject. They are written by experts, and have been translated into more than 45 different languages.

The series began in 1995, and now covers a wide variety of topics in every discipline. The VSI library now contains over 500 volumes—a Very Short Introduction to everything from Psychology and Philosophy of Science to American History and Relativity—and continues to grow in every subject area.

Very Short Introductions available now:

Available soon:

For more information visit our website

www.oup.com/vsi/

Storm Dunlop

WEATHER

A Very Short Introduction

OXFORD
UNIVERSITY PRESS

OXFORD
UNIVERSITY PRESS

Great Clarendon Street, Oxford, OX2 6DP,
United Kingdom

Oxford University Press is a department of the University of Oxford.
It furthers the University's objective of excellence in research, scholarship,
and education by publishing worldwide. Oxford is a registered trade mark of
Oxford University Press in the UK and in certain other countries

© Storm Dunlop 2017

The moral rights of the author have been asserted

First edition published in 2017
Impression: 1

Published in the United States of America by Oxford University Press
198 Madison Avenue, New York, NY 10016, United States of America

British Library Cataloguing in Publication Data
Data available

Library of Congress Control Number: 2016952143

ISBN 978-0-19-957131-4

Printed in Great Britain by
Ashford Colour Press Ltd, Gosport, Hampshire

Links to third party websites are provided by Oxford in good faith and
for information only. Oxford disclaims any responsibility for the materials
contained in any third party website referenced in this work.

Contents

Preface

The weather ultimately affects everyone on Earth, and everything that we try (or hope) to do, whether related to work or leisure. But the Earth's weather systems are extremely complex, and conditions and events may have an affect 'half a world away'. This book aims to explain some of the mechanisms that are at work, and why specific conditions at a particular location may be highly changeable or persist for long periods of time.

There is a famous quotation, attributed to Mark Twain—or perhaps his editor—that runs: 'Climate is what we expect, weather is what we get.' This book only briefly touches on matters of climate: the long-term weather conditions that prevail over a particular area and which are governed by latitude, the proximity to maritime regions, altitude, and similar factors. It is climate that largely determines the type of agriculture that may be carried out in a particular area, governed as that is by the overall prevailing temperatures, the nature and timing of seasons (including dry and rainy seasons), and the variations that may take place from year to year and over a longer term. The actual weather that occurs does, of course, greatly affect the success, or otherwise, of all agriculture and horticulture.

Neither does this book do more than mention the all-important matters of global warming and climate change. It may be said that both of these topics are accepted as real and definitely occurring by the vast majority of meteorologists and climatologists and, as such, are matters of great concern. However, the broader issues that these subjects raise are not covered here. But global warming and climate change are fully expected to create changes in the weather patterns experienced all over the globe. It is the mechanisms behind the actual weather systems that will accompany the expected changes that are covered in this work.

List of illustrations

Chapter 1
The atmosphere

Anyone who has ever taken the mountain railway to the highest station in Europe, at the Jungfraujoch (altitude 3454 m) in the Swiss Alps, will have seen unprepared tourists shivering even under the brilliant sunshine and deep blue skies of midsummer, or tottering in high heels in the tunnel cut through the ice of the glacier. They had simply forgotten (or had never realized) that the higher you go, the colder it gets—at least in the lowest layer of the atmosphere. The change in temperature with altitude is known as the lapse rate, and in the lowest layer, the troposphere (the 'sphere of change' in which most significant weather takes place), it averages about 0.65 deg. C per 100 metres (see Box 1).

Box 1 Temperatures and differences in temperature

To prevent confusion, meteorologists show actual temperatures by the use of the degree symbol (e.g. 20 °C) and differences in temperature by the abbreviation 'deg.' (e.g. 5 deg. C). Note that temperatures are also sometimes expressed in kelvins, named after William Thomson, 1st Baron Kelvin (1824–1907), the physicist, who first identified the need for an absolute thermometric scale. The scale is measured from absolute zero, at which all molecular motion ceases (–273.16 °C). A kelvin is a unit of heat, so temperatures given on the Kelvin scale (such as 273 K) do not use the degree symbol.

If those tourists started their trip at Interlaken (altitude 568 m), the air temperature would have dropped by at least 19 deg. C during their climb.

Temperature changes with altitude

Why does temperature generally decline with height? It is all related to air pressure (see Box 2). Blaise Pascal, the French mathematician and philosopher (1623–62), was the first to prove that atmospheric pressure decreased with height—a suggestion

Box 2 Measuring and charting pressure

Atmospheric pressure is measured by various forms of barometer. The original form of barometer consisted of a glass tube, sealed at the upper end and filled with mercury. Originally pressure was specified by the height of the mercury column, which in English-speaking countries was usually quoted in inches of mercury (in Hg). A unit known as the millibar (mb) is frequently used nowadays, where one millibar is nominally one-thousandth of average sea-level pressure (1 bar). (In fact, the average sea-level pressure has been defined as 1013.2 mb.) Pressure should really be measured in pascals (Pa) or kilopascals (kPa), which are rigorously defined scientifically, but for convenience, meteorologists generally use another unit, the hectopascal (hPa), where one hectopascal (100 Pa) is identical to one millibar (1 hPA ≡ 1 mb).

Pressure is indicated on meteorological surface charts by means of isobars, lines joining locations with equal atmospheric pressure. The charts most commonly seen by the general public display surface pressure, but for forecasting purposes somewhat similar charts are drawn for various levels in the atmosphere. These actually indicate the height of a particular pressure surface (that is, the height at which a specific pressure occurs).

first made by the inventor of the barometer, Evangelista Torricelli (1608–47), who was a pupil of Galileo Galilei (1564–1642). Galileo himself had devised a thermoscope, the precursor of a true thermometer, but this was subject to fluctuations in atmospheric pressure, prompting Torricelli to devise an instrument to accurately measure pressure. In 1648, Pascal persuaded his brother-in-law Florin Perier, with others, to carry components of a barometer and a supply of mercury to the top of the Puy de Dôme (1485 m), making measurements on the ascent and descent, and comparing the readings with a barometer kept at the base of the mountain. The results confirmed the decrease in pressure with height. Pascal himself subsequently carried a barometer to the top of a 50m church bell-tower in Paris and found an identical (but much smaller) effect. Centuries later the SI (Système Internationale) unit of pressure was named in Pascal's honour. It was the famous mathematician and astronomer Edmond Halley (1656–1742) who, in 1686, first developed a mathematical formula relating pressure and altitude.

The change in pressure is directly responsible for the change in temperature, because a decrease in pressure causes a parcel of air—meteorologists often speak of 'parcels' or 'packets' of air—to expand and cool. (Conversely, of course, an increase in pressure causes the air to be compressed and to become warmer.) It might therefore be expected that the temperature of the atmosphere would decrease evenly with increasing altitude as the pressure declines until it reaches interplanetary space, where there is an almost perfect vacuum. Other factors, however, come into play. There is indeed an overall decrease in the lowermost region (the troposphere), but at a certain altitude the temperature steadies, may remain essentially constant for some kilometres, and then begins to increase. Technically, any change in the lapse rate from a decrease or increase is known as an inversion, whether the rate becomes zero or changes sign—from positive to negative or the reverse. However, the term is commonly taken (by both meteorologists and the general public) to imply just an increase in temperature. Inversions may occur at almost any height in the troposphere (and indeed limit the growth of certain clouds), but the inversion at the top of the troposphere is a major feature, always present in the atmosphere.

Here, the inversion defines the boundary (the tropopause) between the troposphere and the overlying stratosphere. (In technical terms, the tropopause is defined as the level at which the lapse rate decreases to 2 deg. C km^{-1} or less, provided that in the next 2 km the rate does not exceed this same value of 2 deg. C km^{-1}.) Although the altitude of the tropopause is variable and may exhibit abrupt breaks and steps, it may be envisaged as generally lying at altitudes of about 16–18 km at the equator, and 9.5–7.5 km, or even less, at the poles, where it is often indistinct, especially in winter. At latitude 45° N or S, the height of the tropopause is approximately 12 km in summer and 10 km in winter. At the equator, the tropopause varies little in its height. Over the high, frigid, Antarctic plateau with an altitude of some 3000 metres, the tropopause is essentially at ground level. A typical temperature at the tropopause is about 218 K (−55 °C). (The various lapse rates in the troposphere and their significance are discussed in more detail in Chapter 4.)

The layers in the atmosphere

The two distinct, lowermost layers of the atmosphere were discovered and named by the French scientist Teisserenc de Bort (1855–1913) through the use of instrumented balloons. He believed (incorrectly) that convective mixing was absent in the higher layer and that therefore gases would separate and become stratified: hence the name 'stratosphere'. Within it, the temperature increases primarily because of the absorption of solar ultraviolet radiation by ozone, O_3, itself created by chemical reactions caused by the radiation. It is this 'ozone layer' that protects the surface from harmful ultraviolet radiation and which has been partially destroyed by man-made chemicals, notably the

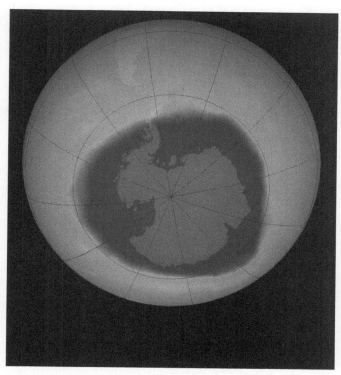

1. **The largest ozone hole ever recorded (October 2015), showing complete depletion of ozone (light tint) over much of Antarctica.**

classes of compounds known as chlorofluorocarbons (CFCs) and hydrochlorofluorocarbons (HCFCs), most of which were banned by the international agreement known as the Montreal Protocol of 1987. The strongest ozone hole develops over Antarctica (see Figure 1), where the polar vortex of strong winds at altitude surrounds the polar region and tends to isolate the area from interactions with the rest of the atmosphere, which would otherwise tend to destroy the ozone hole. A smaller reduction in the ozone layer has been detected over the Arctic, where

2. Polar stratospheric clouds above lower cirrus clouds, photographed from Kirkuna, Sweden.

conditions are less favourable for a partial isolation of the polar region from the general atmospheric flow. Ozone becomes depleted in spring, when sunlight returns to the polar regions and is able to initiate photochemical reactions that occur on the surfaces of ice particles in polar stratospheric clouds (Figure 2), resulting in the destruction of ozone. Although there are concerns about some, previously unconsidered, chemicals not covered by the Montreal Protocol, which appear to be active, in general the international action against harmful chemicals has proved effective, and the ozone holes are showing signs of decreasing in area and severity. It has recently been found, for example, that the area of the Antarctic ozone hole (in September 2015) was 4 million square kilometres smaller than in the year 2000. However, the same study also found that major volcanic eruptions can have an effect, by ejecting large quantities of sulphur dioxide gas (SO_2) into the atmosphere, which combines with water molecules to produce droplets of sulphuric acid (H_2SO_4), which act as nuclei for the formation of the particles in polar stratospheric clouds. This is thought to be the reason for the largest Antarctic hole ever observed (in October 2015), following the eruption of the Calbuco volcano in Chile.

The concentration of ozone in the stratosphere is subject to fluctuations, quite apart from the major ozone holes. It has recently been found that solar events with greatly enhanced proton fluxes (known as solar proton events, SPEs) trigger a decrease in ozone in the ozone layer and an increase in its occurrence in the upper troposphere.

Because of the inversion formed by the tropopause, conditions above that height have only an indirect influence on the weather experienced at the surface. The stratosphere is very dry and clouds are rare. Apart from polar stratospheric clouds (PSC), which, depending on their exact type, may consist of ice or more exotic chemicals (such as nitric acid, sulphuric acid, or nitric acid trihydrate), the only clouds found in the stratosphere are thin wisps of ice-crystal cirrus (often associated with jet streams, see Chapter 3), and the 'overshooting tops' of giant cumulonimbus clouds, where the convection is so vigorous that it is able to break through the inversion at the tropopause into the lowermost stratosphere. When they consist of pure water ice particles, polar stratospheric clouds are sometimes visible as brilliantly hued nacreous, or 'mother-of-pearl' clouds. However, the principal ozone depletion occurs with the particles that have a more complex chemical composition and show little colour.

It has recently been established that regions of the upper troposphere and lower stratosphere may be free from clouds but still contain vast quantities of minute ice particles. These pose a previously unexpected hazard to high-flying aircraft, because they are currently undetectable and exist in regions where ice was not thought to occur. They pose no particular threat to most of an aircraft but melt within a hot engine, where they act to attract more crystals that may build up into large masses of ice that may break off and either damage the engine or cause it to shut down. These minute ice crystals arise from large quantities of water vapour that are lifted to great heights by some unknown mechanism displayed by decaying (rather than active) cumulonimbus clouds.

There are usually major breaks and changes in level of the tropopause near the location of high-speed winds known as jet streams (discussed in Chapters 3 and 4). Jet streams themselves do have a direct influence on the motion of weather systems in the troposphere, particularly depressions, and the resultant changes in weather at the surface.

The stratospheric temperature rises to around 270 K (about 0 °C) at an altitude of some 50 km, where there is another reversal of the lapse rate, which once again starts to decrease with increasing height. This boundary forms the top of the stratosphere and is known as the stratopause. In the overlying layer, the mesosphere, the temperature continues to decline with height, reaching the atmospheric minimum which may be as low as 110–173 K (–163 to –100 °C) at the mesopause. This layer is the coldest region on Earth, and recent studies have revealed that the accepted picture of the changes in temperature with height shown in Figure 3 is incomplete. There are actually two heights at which temperature minima occur: at an altitude of 86 ± 3 km and also at about 100 ± 3 km. In summer, there is up-welling near the summer pole (and corresponding subsidence at the winter pole). The up-welling is accompanied by expansion and cooling of the air, with the result that the lowest temperatures occur at the higher summer mesopause. The subsidence at the winter pole naturally results in the air becoming compressed and heated, with the resultant lower mesopause and higher temperature.

Beyond the mesopause, in the layer known as the thermosphere, temperature increases continuously with altitude, extending into interplanetary space. In this region the concept of temperature begins to lose its meaning, because although atoms and molecules have extremely high velocities, which would normally imply high temperatures, the actual density is exceptionally low. A 'thermometer' of any sort (or any other object) within this region is struck by so few particles that they have little effect in raising the object's temperature. The region beyond 200–700 km is sometimes known as the exosphere, from which atoms or molecules may reach escape velocity and cease to be bound by Earth's gravitational attraction. The lower boundary applies when there is intense solar activity.

Over the years, many scientists have attempted to find a link between solar activity (usually defined by the number of sunspots, the frequency of which generally follows an approximately eleven-year cycle) and the weather. Although it is now known that solar activity displays an approximately twenty-two-year, magnetic reversal cycle, and some have claimed to detect an eighty-year cycle in solar activity, no scientifically or statistically adequate link was ever found between sunspot numbers and weather at the Earth's surface. Only in the last decade has a possible mechanism been discovered through which solar activity influences the weather. This induces major north–south excursions and 'blocking' of the polar jet stream in the northern hemisphere. (Jet streams and blocking are discussed in more detail in Chapter 3.)

Although not directly related to weather near the surface, it may be noted that the ionosphere, where solar ultraviolet and X-ray radiation strip electrons from atoms and molecules, lies between $c.60$–70 km and 1000 km or more, encompassing the upper mesosphere and thermosphere. It reflects certain radio wavelengths back towards the ground and simultaneously prevents similar wavelengths from reaching Earth's surface from space. Aurorae, which are related to solar activity, occur within the ionosphere, when energetic electrons and protons from the solar wind bombard atmospheric oxygen and nitrogen atoms and molecules, either ionizing them or raising them to an excited state, from which they return with the emission of visible light.

For practical purposes, particularly for aviation (such as determining the performance of aircraft at different altitudes) and for the calibration of scientific instruments, the change in temperature, pressure, and density with height is represented by what is termed a standard atmosphere. This is a hypothetical, idealized distribution, using certain assumptions about the physical properties of air, the pressure at sea level, and the lapse rates and temperatures at specific altitudes. (It assumes that the temperature at the tropopause is –56.5 °C, for example.) The International Civil Aviation Organization (ICAO) standard atmosphere is shown to 100 km in Figure 3, and the American extension to this standard (the US Standard Atmosphere) is used to 500 km.

Atmospheric composition

In the lower atmosphere, the air consists of approximately 78 per cent nitrogen and 21 per cent oxygen by volume (the exact abundances are shown in Table 1). This shows that, taken together, all the other gases—including the carbon dioxide and methane that are largely responsible for global warming—amount to less than 1 per cent. These figures apply to completely dry air, but there is one extremely important component that is present in variable amounts, and this is water vapour, which may vary between zero and approximately 4 per cent. The properties of water are so important in determining the weather that we will discuss them in some detail in Chapter 4.

With the exception of water vapour, carbon dioxide, and ozone, which are present in variable amounts, the relative composition of the atmosphere remains effectively constant, in what is sometimes called the homosphere, up to an altitude of about 85–100 km (i.e. to about the altitude of the mesopause). This region is also known as the turbosphere, in which mixing occurs by turbulence.

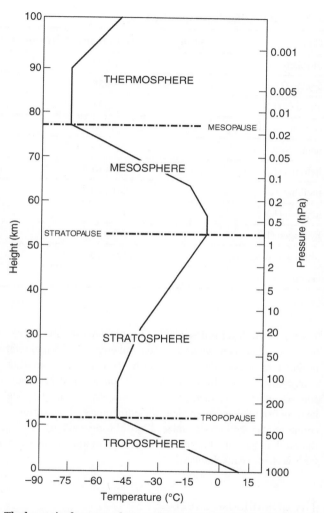

3. The layers in the atmosphere and the temperature profile as used in the International Civil Aviation Organization (ICAO) standard atmosphere.

Table 1 Atmospheric composition

Gas	Atmospheric abundance (% by volume)
nitrogen (N_2)	78.09
oxygen (O_2)	20.95
argon (Ar)	0.94
carbon dioxide (CO_2)	~0.03
neon (Ne)	1.8×10^{-3}
helium (He)	5.2×10^{-4}
methane (CH_4)	2.0×10^{-4}
krypton (K)	1.0×10^{-4}
hydrogen (H)	5.0×10^{-5}
nitrous oxide (N_2O)	5.0×10^{-5}
xenon (Xe)	8.0×10^{-6}

The top of this layer is described as the turbopause. Above this height, in what is known as the heterosphere, the types and relative proportions of gases do vary, mainly because of the dissociation of oxygen molecules by ultraviolet radiation, and also because convection—the overturning of a gas or liquid, normally through heating from below, and which efficiently mixes all the various components together—is replaced by diffusion—mixing through the random motion of individual molecules or atoms—which is a far less effective mechanism.

The greenhouse effect

The various processes that give rise to weather on Earth are ultimately driven by energy from the Sun. (For our purposes we may ignore the small contribution of energy in the form of heat from the Earth's interior arising from the decay of radioactive elements present when the planet was originally formed.)

A planetary greenhouse effect was first suggested (by Joseph Fourier) as far back as 1824. It was more fully described by the famous Swedish scientist and Nobel Prize winner Svante Arrhenius (1859–1927), in 1896. As early as 1917, Alexander Graham Bell (1847–1922) pointed out that burning fossil fuels would create a greenhouse effect. He also advocated the use of alternative energy sources, and specifically solar energy.

Any discussion of global warming or climate change is likely to mention the terms 'greenhouse effect' or 'greenhouse gases'. Although convenient, but like certain other terms in popular usage, use of the word 'greenhouse' here is not, strictly speaking, correct. The inside of a greenhouse or conservatory of any sort is usually much warmer than the outside air. This applies whether it is a small domestic greenhouse or the giant domes at the Eden Project in Cornwall in England, within which the ecological communities (the 'biomes') of a tropical rainforest and Mediterranean regions are reproduced. Sunlight heats the soil, which in turn heats the air. But the glass (or plastic) simply prevents convection, which would otherwise mix the hot inside air with cooler air outside. The giant domes of the Eden Project require complex ventilation systems to maintain their required temperature and humidity ranges.

In the case of an atmosphere, the warming mechanism is rather different. Here, the planetary greenhouse effect certainly causes the surface temperature of the planet or any planetary satellite to be higher than it would be without an atmosphere. The Earth's average surface temperature is about 14 °C, but without an atmosphere it would be a chilly –18 °C, far too cold for complex life forms to arise and thrive. The temperature difference occurs because of the way in which the various gases absorb and emit different wavelengths of infrared radiation.

The Sun emits radiation over practically the whole electromagnetic spectrum, from short-wavelength gamma- and X-rays, and ultraviolet (UV) radiation to long-wavelength radio emissions. Figure 4 shows how the opacity of the atmosphere varies at different wavelengths, preventing much of the Sun's radiation from reaching the surface. Extremely short-wave gamma-rays, X-rays, and UV radiation are absorbed by the upper atmosphere where, as we have seen, the UV radiation acts on O_2 molecules to create the ozone layer in the stratosphere. There is a narrow window that allows most visible light to reach the surface (with some slight absorption). Other, longer wavelengths in the infrared region are blocked to a greater or lesser degree, until we reach the wide window in the radio region where signals reach the ground unimpeded. The very longest radio wavelengths are again blocked completely.

Carbon dioxide and water vapour, in particular, are primarily responsible for the absorption in the infrared region. Some of the incoming radiation is reflected back into space by clouds, aerosols (described shortly), and the surface, but most of the radiation that does reach the surface is absorbed, raising its temperature. Now the ground itself emits radiation towards space, but this is at longer wavelengths, some of which are blocked by the 'greenhouse gases' in the atmosphere. They, in turn, re-emit the radiation, some being directly radiated to space, but the remainder being returned towards the surface, raising its temperature even more. So the atmosphere and the greenhouse gases act as a form of partial 'one-way' blanket, trapping heat and thus raising the overall average temperature of the Earth. Taken as a whole, of course, the incoming energy from the Sun (an average of 342 Wm^{-2}) is balanced by an equal amount that is returned to space and this mean annual energy balance and its various components are shown in Figure 5. The way in which the incoming solar radiation creates both vertical and horizontal motion in the atmosphere is discussed in Chapter 2.

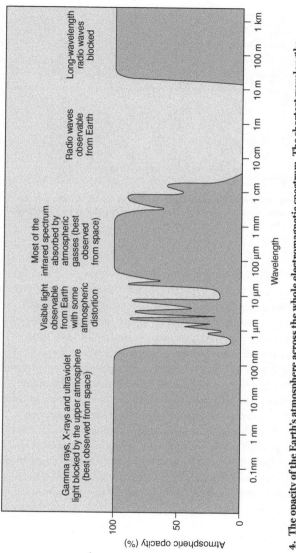

4. The opacity of the Earth's atmosphere across the whole electromagnetic spectrum. The shortest wavelengths (gamma-rays, X-rays, and most ultraviolet light) are blocked by the atmosphere. There is an almost complete window in the visible range, blocking at most infrared wavelengths, and a large radio window.

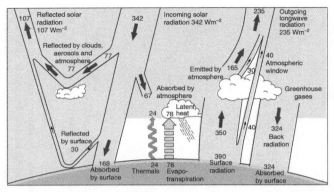

5. The complex mean annual energy budget for the Earth and its atmosphere. Note the large amount of energy (324 Wm^{-2}), returned to the surface by greenhouse gases.

The absorption spectrum shown in Figure 4 incorporates the effects of those atmospheric components that are always present (including water vapour), but does not take into account the varying concentrations of solid or liquid particles that may be present at different times. Aerosols (minute liquid or solid particles that remain suspended in the air) not only create pollution hazes when close to the surface, but have a significant effect on planetary temperatures when present in the upper atmosphere. The eruption of Mount Pinatubo in June 1991, for example, ejected an estimated 14 million tonnes of sulphur dioxide gas (SO_2) into the stratosphere, where it combined with water vapour to produce droplets of sulphuric acid (H_2SO_4). The tiny droplets blocked solar radiation, causing a short-lived cooling of the entire Earth in the early 1990s, which initially (for some months) amounted to about 0.5 deg. C. It has recently been established that aerosols interact with atmospheric gases. In the case of carbon monoxide and methane, they increase the warming effect (relative to carbon dioxide's warming potential), whereas nitrogen oxides (produced naturally by lightning and in large quantities by vehicle exhausts) have a major cooling effect.

The blue of the sky arises because oxygen and nitrogen molecules preferentially scatter violet and blue light in all directions and have little or no effect on longer wavelengths. The sky therefore appears blue to human eyes (which are unable to detect violet light). The yellow, orange, and red colours at sunrise and sunset arise mainly because the shorter wavelengths have been scattered aside before they reach the observer. Water vapour is naturally concentrated towards the surface, where the molecules of H_2O tend to scatter light of all wavelengths, giving a milky, lighter blue tint to the sky, except in polar or desert conditions when humidity is extremely low. At high altitudes—such as those experienced by those tourists at the Jungfraujoch—there is little water vapour, so the sky appears a very deep shade of blue.

At high mountain altitudes the air is often very cold and dry and in ascending to a peak one frequently passes through a layer of cloud, only to find clear skies above. When clouds cover the highest peaks, it is usually so cold that any precipitation is in the form of snow. Major astronomical observatories are located on the top of mountain peaks, because they are often 'above the weather' and the optical conditions are greatly improved by the lack of pollution—which tends to be confined to the lowest layer of the atmosphere—the low concentration of water vapour, and the fact that light from cosmic objects has a shorter path through the distortions imposed by turbulence in the atmosphere. The low concentration of water vapour is particularly important for astronomical observations at millimetre and sub-millimetre wavelengths, which are otherwise completely blocked or severely attenuated. The Atacama Large Millimetre Array (ALMA), for example (Figure 6), is located on the plateau of Chajnantor in Chile at an altitude of no less than 5058 m, at which height

6. The Atacama Compact Array, consisting of sixteen closely spaced units of the fifty antennae in the overall Atacama Large Millimeter Array (ALMA).

workers on the radio telescopes require supplementary oxygen. Overall operation of the array is carried out from the much lower Support Facility at an altitude of 2900 m.

Chapter 2
The circulation of the atmosphere

The first serious attempt to explain the overall circulation of the atmosphere was made in 1686 by the famous mathematician and astronomer Edmond Halley (1646–1742), who linked the distribution of solar heating over the Earth's surface with the winds. In 1676, Halley visited St Helena in the South Atlantic to catalogue stars in the southern hemisphere. During his outward and return journeys he observed the pattern of winds on both sides of the equator, and subsequently (in 1686) published the first meteorological map (Figure 7), showing winds in the tropics. He realized that heated air rising in the equatorial regions drew in cooler air from the surrounding area, and thus explained the pattern of winds that he experienced. Although his explanation of the easterly trade winds was faulty, his concept of thermal convection was basically correct and, as a result, he has sometimes been called the 'Father of Dynamic Meteorology'. ('Dynamic' or 'Dynamical Meteorology' is the branch of meteorology that studies the processes creating motion in the atmosphere.)

In 1735 George Hadley (1685–1768), the lawyer and amateur meteorologist—who is often confused with his rather more famous brother John Hadley, the astronomer and optical-instrument maker, who invented the sextant—expanded Halley's work. George Hadley proposed that the contrast in temperature between the tropics and the poles sets up two circulation cells, one on each side

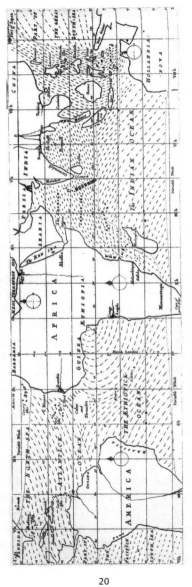

7. Edmond Halley's 1686 map of the winds in the tropics.

of the equator. In each cell, air was thought to rise over the equatorial zone, flow towards the pole, where it sank, subsequently flowing back to the tropics at a low level. Although we now know that this simple model with a north–south (meridional) circulation was incomplete, Hadley did correctly surmise that the north-easterly and south-easterly direction of the trade winds that diverged from a north–south line and converged at the equator arose because of the rotation of the Earth. (The circulation of the lower atmosphere of Venus does exhibit a single Hadley cell in each hemisphere between the equator and latitudes of approximately 60–70° N and S.)

The 'trade' winds were not so named because they had any relation to commerce—important though they were for ships crossing the oceans. The name arises from the obsolete term 'blow trade' (where 'trade' originally meant 'course' or 'path' and came to be understood to imply 'customary' or 'habitual'). It was thus used to indicate that over certain regions of the Earth the winds blow essentially continuously and consistently from a particular direction.

Although Hadley's model explained, after a fashion, the north-easterly and south-easterly trade winds in the zones on either side of the equator (described in more detail in Chapter 6), it fell down badly in accounting for the persistent bands of predominantly westerly winds that occurred further towards the poles. But he (like Halley before him) was correct in assuming that temperature (and thus pressure) differences were the driving force.

The insolation, or amount of energy received from the Sun by the Earth's surface, is very strongly dependent on latitude. On average, taken over the year, there is an energy excess in the region between the equator and latitudes of approximately 40° N and 40° S, and a deficit over all latitudes closer to the poles. This does lead to a form of cellular circulation. Where warm moist air is rising near the equator, there is a low-pressure region (a 'thermal low') known historically to sailors as the Doldrums, where winds are light and variable, because the main motion is vertical, rather than horizontal. At altitude, the air flows away to north and south, descending at about 30° N and 30° S to create quasi-permanent areas of high pressure (known as the subtropical anticyclones). Because the air warms as it descends, such an anticyclone is termed a 'warm high'. Its humidity becomes extremely low, and on reaching low levels, the stream of hot, dry air divides, some flowing back towards the equator as the trade winds, and the remainder flowing towards higher latitudes. The world's major hot deserts are found near or on the equatorial side of the subtropical anticyclones: in particular the Sahara, the Arabian Desert, and the desert lands of North America in the northern hemisphere; and the central Australian and the Kalahari Deserts in the south.

A somewhat similar circulation occurs towards the poles. The air at latitudes of approximately 60° N and 60° S is sufficiently warm and moist to convect, rise, and move towards the poles. At the poles where temperatures are low, the surface air is already cold and dense, and the upper air cools and descends. The result is a shallow, high-pressure region (a cold anticyclone or 'cold high') from which air flows out at the surface towards lower latitudes, eventually replacing the air that rose at around latitude 60°, completing the circulation cell. This cold air encounters the warmer air flowing out of the subtropical anticyclones along an extremely important frontal zone, known as the Polar Front, which will be described in more detail in Chapter 3. Here, some of the warm air from the subtropics is dragged into the circulation of the polar cell, and some rises and flows back towards lower latitudes.

Instead of a single cell in each hemisphere, there are thus three cells: the cell nearest the equator, which corresponds most closely to Hadley's original idea (and is called the Hadley cell in his honour); the polar cell; and an intermediate cell over the middle

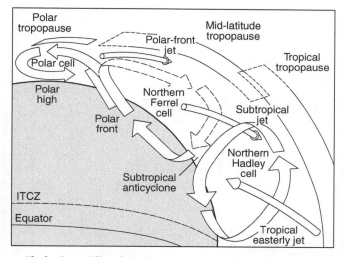

8. **The basic meridional circulation, corresponding to the northern summer and showing the three primary circulation cells (the polar, Ferrel, and Hadley cells).**

latitudes, known as the Ferrel cell after an American meteorologist, William Ferrel (1817–91). Air descending at the subtropical anticyclones around 30° N and 30° S and the air rising at around 60° N and 60° S thus forces circulation in the intermediate Ferrel cell. The Hadley and polar cells are regarded as 'direct' cells, driven by specific temperature and pressure gradients, and the Ferrel cell is considered to be 'indirect', completing the circulation established by the two direct cells. A simplified representation of this meridional circulation—ignoring for the moment the effects of the Earth's rotation—is shown in Figure 8.

This simplified explanation serves our purpose here, but the actual situation is far more complex. There is not, for example, a single Hadley cell extending all the way round the Earth, nor continuous bands forming the subtropical anticyclones. Instead there are a series of individual circulation cells, limited in

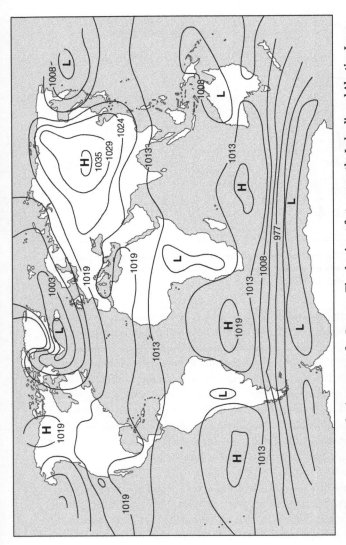

9a. The average atmospheric pressure for January. The dominant features are the Icelandic and Aleutian Lows, the Siberian High, and the three semi-permanent high-pressure regions over the Southern Ocean.

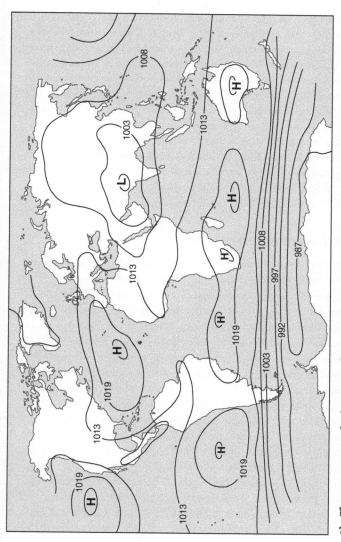

9b. The average atmospheric pressure for June. The dominant features are now the Azores/Bermuda High, the North Pacific High, and the Asian Low. The high-pressure regions in the southern hemisphere are relatively unchanged.

longitude and a number of high-pressure centres. This is suggested, although not explicitly indicated, in Figures 9 and 12 (showing the mean January and June distribution of pressure and wind directions, respectively). Neither is the flow of air as well defined as descriptions and diagrams may suggest. Not all of the air from the Hadley cells descends at the subtropical anticyclones, but some continues towards the poles at altitude. Similarly, although the air that is initially close to the surface in the Ferrel cell rises along the Polar Front and immediately splits at altitude with some returning towards the tropics and some continuing towards the poles, a small fraction descends behind the Front itself. The boundaries between the cells are, of course, subject to considerable variation in latitude, particularly with the changing seasons, and there is also an interchange of air between them.

Global pressure patterns

The surface winds arise from the uneven distribution of heating (and hence pressure) around the globe. The distribution of high- and low-pressure regions naturally varies throughout the year, tending to shift north and south with the seasons, following maximum insolation. The average distribution of surface pressure for January and July is shown in Figure 9. The most striking differences between the two patterns are the major change in the pressure over Siberia from high pressure in winter (the Siberian High) to low pressure in summer (the Asian Low, centred further to the south); and the low-pressure regions that occur in winter over the northern Atlantic and Pacific Oceans (the Icelandic Low and Aleutian Low, respectively). These semi-permanent features are known as 'centres of action'. The high-pressure centres tend to be relatively constant, but some of the low-pressure centres are more fugitive, only existing because depression systems frequently cross the areas concerned. During summer, centres of action such as the Azores High (often known as the Bermuda High) and the North Pacific High both strengthen as the Icelandic and Aleutian Lows weaken. Because of the predominance of oceans in the southern hemisphere, changes there are less dramatic, although the three high-pressure regions present in January tend to expand and merge to encircle the globe during the southern winter. In the southern summer there are significant lows over northern Australia, over central and southern Africa, and around the Antarctic continent, although the last tends to persist throughout the year.

The circulation of surface winds is largely determined by the distribution of surface pressure, but here the rotation of the Earth plays an extremely important part. The difference in atmospheric pressure between high- and low-pressure areas creates a pressure gradient acting from the high towards the centre of the low. It might be expected that air would flow directly towards the centre of the low. In fact, the rotation of the Earth is the principal cause for the flow to be deflected from the straight-line path otherwise created by the pressure gradient.

The Coriolis effect

When viewed in a rotating frame of reference (such as from the surface of the Earth), there is an apparent tendency for any moving object to deviate from a straight path. This is known as the Coriolis effect and applies to material objects such as rockets or artillery shells as well as to parcels of air. In meteorology, the most significant part of this Coriolis acceleration is the horizontal component that acts parallel to the surface of the Earth, and which is generally known as the Coriolis force. A full description is beyond the scope of this book, but a simplified explanation will serve our purpose.

Because of the Earth's rotation, a point on the equator (and a stationary parcel of air above it) is carried eastwards at a rate of 40,074 km in twenty-four hours (at a velocity of approximately 1670 kmh^{-1}), whereas a point at one of the poles has no horizontal motion but merely rotates once in twenty-four hours. If, for the sake of simplicity, we assume that, under the influence of a pressure gradient, an equatorial parcel of air moves directly north or south, it retains its eastward velocity but the surface now beneath it is moving at a lower velocity. The parcel will move east, relative to the lines of longitude. It is no longer flowing directly towards the low-pressure centre. At the poles, by contrast, a parcel of air has a slow rotation about the Earth's axis. If it begins to move towards the equator it will move over a surface that is moving more rapidly towards the east. Both parcels will appear to turn towards the right in the northern hemisphere and towards the left in the southern.

These illustrative examples might suggest that the deflection, the Coriolis force, exists at the equator. In fact, the magnitude of the force is proportional to the sine of the latitude, so its magnitude is actually zero at the equator (sin 0° = 0), and reaches a maximum at the poles (sin 90° = 1). In addition, it is proportional to the horizontal velocity of the air, and always acts at right angles to the direction of air motion (to the right in the northern hemisphere and to the left in the southern).

It has been found that in the 'free atmosphere', which is usually taken to be above an altitude of 500–1000 metres—i.e. above the level affected by surface friction—the wind blows at right angles to the pressure gradient and thus along the isobars. This implies that the pressure gradient and the Coriolis force are acting in opposite directions and exactly balance one another. This theoretical wind is known as the geostrophic wind, and for a given pressure gradient, its velocity is inversely dependent on the sine of the latitude, thus decreasing towards the poles. It closely approximates the wind observed in the free atmosphere, except close to the equator, where the Coriolis force approaches zero. Because pressure centres are very seldom stationary in either latitude or longitude, and the central pressures also vary with time, the velocity of the air is continually changing as a result of alterations to the balance between the various forces.

Because the freely flowing (geostrophic) wind blows along the isobars, it circulates around centres of high and low pressure. An additional force, acting towards the centre and known as the centripetal force (or centripetal acceleration), keeps the air moving along the curved path. Air moving in this way is known as the gradient wind. The strength of the force depends on whether the air is circulating around a region of low or high pressure. It tends to decrease the velocity of the air around a low-pressure centre, and increase it around a high-pressure area. In most cases, this additional factor may be ignored and it is only of major significance in tropical cyclones near the equator, where the Coriolis force is low or non-existent, or where extremely low pressures are encountered in tornadoes.

This 'simplified explanation' may appear moderately complicated, but the important points to note are that a moving parcel of air will deviate to the right in the northern hemisphere (and to the left in the southern); that the Coriolis force increases with increasing latitude and is greater, the greater the wind speed; and, finally, that freely flowing air (the geostrophic wind) flows along the isobars and around centres of high or low pressure. Friction at the surface (to be described shortly) modifies this behaviour.

In this book, wind speeds are generally given in kilometres per hour (kmh^{-1}), felt to be more familiar to readers than metres per second (ms^{-1}), the unit generally used in meteorology. In aviation and other fields, however, speeds may be given in knots (kt)—nautical miles per hour—where 1 kt = 0.514 ms^{-1} = 1.852 kmh^{-1}. Wind speeds are also often given in the Beaufort scale, introduced by Rear-Admiral Francis Beaufort for use at sea in 1806, adopted by the British Admiralty in 1838, and later adapted for use on land. (The Beaufort scale is given in Appendix A.)

Global winds

We now have an explanation for the direction of the winds forming the global circulation. The air rising at the equator and flowing north and south at altitude will be south-westerly in the northern hemisphere, and north-westerly in the southern. (Wind directions are described below) The air returning at low level towards the equator from the subtropical anticyclones gives rise to the north-easterly trades and south-easterly trades in the northern and southern hemispheres, respectively. There is an important meteorological feature where the trade winds meet, known as the Intertropical Convergence Zone (ITCZ). This is also sometimes called the 'near-equatorial trough'. The ITCZ is often visible on satellite images as significant amounts of cloud, and particularly thunderstorm clusters (Figure 10). It shifts north and south and also changes the path it follows around the Earth with the changing seasons.

It is useful to remember that the wind direction is always described by the point from which it originates. This applies both to winds described in general terms relative to a compass point (easterly, north-westerly, etc.), and to the more specific winds, such as mountain winds, which will be discussed in Chapter 8.

10. The Intertropical Convergence Zone (ITCZ) over the eastern Pacific, as imaged by a meteorological satellite.

The low-level air flowing out of the subtropical anticyclones towards the poles produces the westerly winds that are dominant over the temperate zones. In the northern hemisphere, these westerlies are strongly influenced by the continental land masses and mountain ranges that are present, but in the south, with less land area, they flow almost unimpeded around the Earth and came to be known to sailors as the Roaring Forties. (Occasionally the strong winds further south are imaginatively described as the Fearsome Fifties and the Screaming Sixties.)

In both hemispheres, the air flowing out of the polar highs produces polar easterlies, but these tend to be confined to areas very close to the poles, sometimes reaching approximately 60° N and S. They are relatively weak, however, especially in the Arctic, so the polar easterlies are not as strong or persistent as the trade winds. Over most of the temperate zones the westerlies predominate. A highly simplified representation of the surface circulation in the northern hemisphere is shown in Figure 11.

The overall wind field is, of course, highly complex and varies greatly from year to year. The mean global circulation for January and July is shown in Figure 12. As with the pressure pattern shown in Figure 9, the most striking difference involves the change between the air flowing out from Siberian High in winter and into the Asian Low, located farther south in summer. In effect, the ICTZ and its accompanying low pressure has retreated to a much higher latitude over the Indian subcontinent, and over that region it is sometimes described as the 'monsoon trough'. The resulting reversal of wind directions produces the monsoon winds (from the Arabic word for 'season'), with the wintertime, predominantly north-easterly monsoon over much of Asia being replaced by the south-westerly monsoon in summer, where air from the southern hemisphere's south-east trades has crossed the equator and become the south-west monsoon. Similar changes occur over northern Australia, West Africa, and, to a lesser extent, over the south-western area of the United States. The June

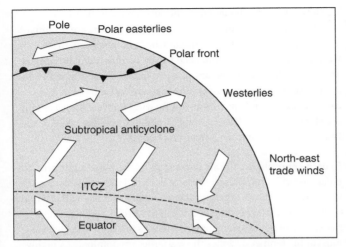

11. A highly simplified diagram showing the pattern of surface winds in the northern hemisphere. The highly variable location of the Polar Front and those of the associated high- and low-pressure systems are not represented.

circulation also reveals how the Azores/Bermuda High in the North Atlantic becomes a significant centre of action during summer. There is a similar, although less distinct, change affecting the North Pacific High.

The westerlies

The circulation in middle latitudes (in the Ferrel cell) is dominated by the zonal, rather than meridional, flow of the westerlies. The location of the Polar Front at the boundary between the Ferrel cell and the polar cell is largely determined by the strength, or otherwise, of the polar vortex (briefly mentioned in connection with the ozone holes). These polar vortices are low-pressure zones near the Earth's poles, in both hemispheres. They lie in the middle and upper troposphere, extending into the lower stratosphere.

They are particularly strong in winter, when the air over the poles becomes extremely cold, because of the lack of sunshine, and the greatest temperature contrast exists between the polar air and the air at lower latitudes, which is prevented from entering the polar region by the strong vortex. In the south the vortex lies beyond 30° S, and is particularly symmetrical about the pole. It is strongest during the southern winter, and weakens only slightly during the summer. In the north, the vortex is more variable, and the strength of the winds varies considerably. When at its strongest the main low-pressure centre lies over the Canadian Arctic (Figure 13, top). If the vortex weakens slightly, a secondary low tends to develop over Siberia. At its weakest, the overall organization breaks down, and multiple cold, low-pressure centres develop. Lobes of cold air push down to lower latitudes (Figure 13, bottom). Generally in winter the westerly winds are particularly strong around the overall, deep low-pressure region and blow anticlockwise around the North Pole). The vortex weakens considerably during the northern summer, and there is then normally a reversal of direction, with moderate easterly winds around a high-pressure centre. There are major latitudinal meanders in the upper flow, most marked in winter, with high-pressure ridges extending north, and low-pressure troughs extending towards the south. These waves are known as long waves or Rossby waves, with wavelengths of thousands of kilometres, which travel eastwards in the overall westerly flow. There are two major semi-permanent troughs of low pressure at about 70° W and 150° E, to leeward of the major topographic barriers of the Rockies and the Tibetan Plateau. (For reasons too complex to discuss here, a pattern of low- and high-pressure areas develops downwind of even low hill or mountain obstructions to the airflow.) There are no such major barriers in the southern hemisphere, where only the southernmost tip of South America extends beyond 40° S, so Rossby waves are less significant. In the north, they have a major effect on the growth and movement of weather systems, as discussed in Chapter 6. The Polar Front jet streams (discussed in Chapter 3) are the high-velocity cores of the Rossby waves.

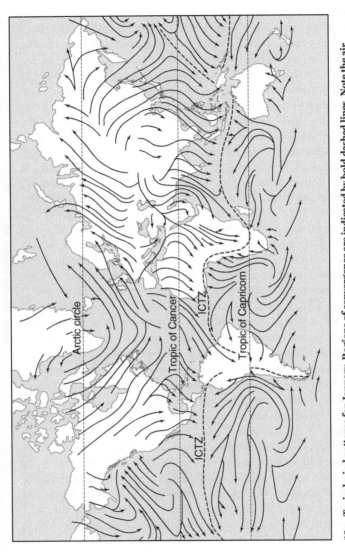

12a. Typical wind patterns for January. Regions of convergence are indicated by bold dashed lines. Note the air flowing out of the Siberian High, and the dip in the Intertropical Convergence Zone (ITCZ) over Africa.

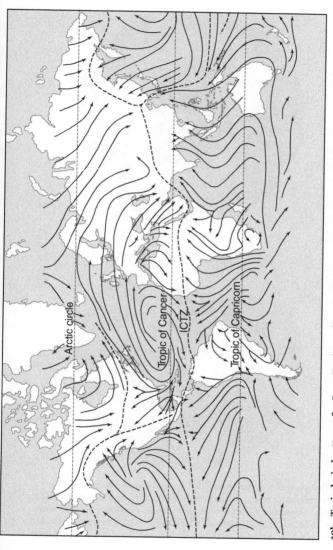

12b. **Typical wind patterns for June.** Regions of convergence are indicated by bold dashed lines. There is a major shift in the location of the Intertropical Convergence Zone (ITCZ) over India and southern Asia, resulting in the reversal of the monsoon winds, and the Azores/Bermuda High has become prominent in the North Atlantic.

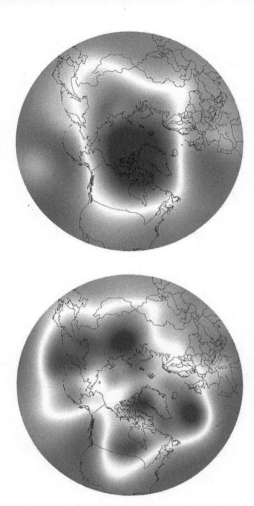

13. The northern polar vortex in a typical strong phase (14–16 November 2013, top), with a main low-pressure centre over northern Canada. When the vortex weakens, it displays multiple centres and lobes (5 January 2014, bottom). The bands of the strongest winds are shown as white, with the darkest regions showing the lowest pressure.

Airflow around high- and low-pressure regions

As we have seen, freely flowing air does not directly follow the pressure gradient from high pressure to low, but instead flows around the pressure centres. The circulation is clockwise around a high-pressure region (an anticyclone or, simply, a 'high') in the northern hemisphere, and anticlockwise around a low-pressure region. The latter is known technically as a cyclone, and less formally as a 'low' or depression.

Close to the surface, however, the wind speed will be reduced through friction and, because the Coriolis force is proportional to the wind speed, it will be reduced. The result is that air flows across the isobars, inwards to the centre of a low, and out from the centre of a high. The amount of friction (and thus the angle at which the wind blows across the isobars) depends on the nature of the surface. Typical values are 10–20° over the sea, and 25–35° over land. This has obvious, important consequences: with convergence in the centre of a low, air cannot accumulate indefinitely, so there must be a compensating ascent of air in the centre with a corresponding divergence at altitude. In a high there is the opposite situation, with divergence at the surface and convergence at some higher level. The direction of the low-pressure centre may be estimated by an observer on the ground by the use of Buys Ballot's Law (see Box 3).

Another result of surface friction is that there is a gradual variation in the wind's direction with height, which, if plotted on a diagram, results in a spiral, known as an Ekman spiral, similar to one that was originally discovered to apply to ocean currents, where the direction varies with depth, increasing with latitude, and may even result in a reversal of flow, typically at depths of some 50 metres, over large regions of the oceans.

The effects of friction, turbulence, and heating from the surface are largely confined to the lowermost layer of the atmosphere,

Box 3 Buys Ballot's Law

There is an approximate rule for finding the position of the centre of low pressure that is creating a wind. In the northern hemisphere, with one's back to the wind, the centre is on the left (right in the southern hemisphere). This is only very approximate, however, because the amount of friction, depending on the surface, will cause the actual centre to lie some 10–35° (or more) forward of that direction. The 'Law' is named after the Dutch meteorologist C. H. D. Buys Ballot (1817–90), who described the effect. He was actually anticipated in this by the American meteorologist William Ferrel, but the rule came to be named after Buys Ballot.

When the wind is flowing from the interior of an anticyclone, the high-pressure centre will, naturally, be on the right in the northern hemisphere, and on the left in the southern, and again shifted by at least 10–35°, but now lying behind the observer.

which is known as the planetary boundary layer (also sometimes called the 'friction layer'). It is generally accepted that this extends to heights of about 500 m over the oceans and 1500 m over land. However, this planetary layer may itself be considered to consist of two separate, individual layers, the surface boundary layer, up to a height of about 10 m, and an overlying Ekman layer. Within the surface boundary layer, wind strength and direction are essentially constant, primarily determined by the roughness and contours of the ground. This thin layer experiences the greatest heating from the ground, and over a very hot surface the lapse rate (change of temperature with height, see Chapter 4) may be considerably greater than lapse rates found higher in the atmosphere. Within the higher, Ekman, layer the wind strength and direction are affected by friction. With increasing height, the effect of friction lessens, and the wind direction changes continuously in an Ekman spiral, eventually becoming geostrophic, flowing along the isobars. The height at which the wind becomes geostrophic defines the top of the Ekman layer and thus the top of the planetary boundary layer.

Ocean currents

The action of the surface winds produces the various ocean currents, although there is also a significant contribution from what is called the thermohaline circulation, where differences in density (created by variations in temperature and salinity)

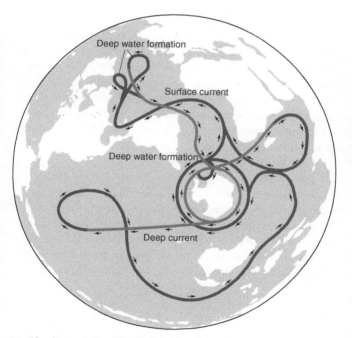

14. The thermohaline circulation (the Great Ocean Conveyor Belt) on a special projection to show the oceans as a single body of water. Cold, dense water forms and descends at two locations in the North Atlantic and in the Weddell Sea off Antarctica.

cause water to sink in the North Atlantic near Greenland and in the Weddell Sea in Antarctica (Figure 14). The dense water in the Great Ocean Conveyor (as it is known) circulates slowly at depth through all the world's oceans and eventually (after 1000 to 1500 years) rises and combines with the surface flow in the North Atlantic to complete the circuit. Although ocean currents do transport heat from low latitudes to higher ones, the atmosphere's contribution is more significant. It is often said that the mild climate of Western Europe, for example, is the result of the warmth arising from the waters of the North Atlantic Drift—a branch of the North Atlantic Current, itself the continuation of the Gulf Stream—but this is technically incorrect. Heat transported by the atmosphere makes a far greater contribution to the mild climate. Computer simulations have shown that if the Rockies did not exist, Rossby waves in the northern circumpolar vortex would be reduced, and the succession of weather systems affecting Western Europe would be greatly modified and it would experience a far colder climate. However, the heat transported by the westerly winds is ultimately derived from the oceanic surface, so the North Atlantic Drift is indirectly responsible for the warmer climate of Western Europe.

The general circulation of the atmosphere and of the ocean currents may seem to be of relevance to climate studies, rather than of significance to the weather that one experiences. In fact, the nature of modern weather forecasting is such that forecasts extending three days or more into the future require an accurate knowledge of conditions over the whole of the Earth at any one time, including a knowledge of sea-surface temperatures.

Chapter 3
Global weather systems

Nearly everyone has been struck, at some time, by the difference as they have moved from the cosseted conditions of a centrally heated or air-conditioned house, office, or shop into the outside air. That outside air may seem hot and humid, hot and dry, damp and cold, or simply freezing cold. Such qualities of the air are often extremely marked and readily apparent. On occasion, vigorous weather systems may create abrupt changes that are detectable even indoors, and less active systems may still produce marked differences overnight or during the course of a day. These changes occur when one air mass replaces another with distinctly different properties.

Air masses

When air remains stationary for some time over a particular region of the Earth, it takes on two specific characteristics (temperature and humidity) that strongly depend upon its precise location. Such volumes of air are known as air masses, and the areas over which they form are known as the source regions (Figure 15). The horizontal extent of air masses is typically hundreds or even thousands of kilometres across. Their depth is strongly dependent on where they form and may range from about 1000 metres to the whole depth of the troposphere. Cold air masses are usually shallow, whereas warmer ones are deeper, simply because heating causes warm air to expand and produce a

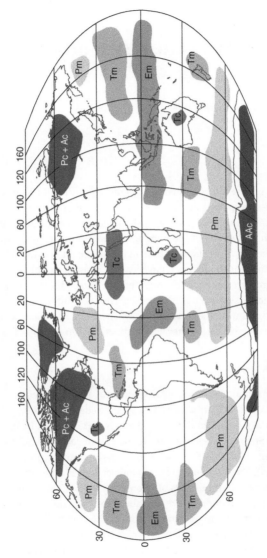

15. **The approximate locations of the primary source regions for different air masses.**

taller column. In a warm air mass, convection acts to equalize temperature and humidity throughout a greater depth of air.

There are various schemes for classifying air masses, but the most common is known as the Bergeron classification. Two broad categories, continental (c) and maritime (m), are used to describe whether the source region is over land or sea, and thus whether the air is normally dry or humid.

It should be noted that many works reverse the order of the two letters (and usually the wording as well) designating each air mass: e.g. 'cA' instead of 'Ac' ('continental arctic' for 'arctic continental'). The convention followed here is that used by the UK Met Office.

Based on temperature, air masses may be of four different types:

- arctic or antarctic (A)
- polar (P)
- tropical (T)
- equatorial (E)

(If it is necessary to differentiate antarctic air from arctic air, the former is designated by 'AA'.)

In combination, the principal types of air mass that are encountered are:

- arctic continental (Ac)   extremely cold and dry
- polar continental (Pc)   cold and dry
- tropical continental (Tc)   hot and dry
- arctic maritime (Am)   extremely cold and humid
- polar maritime (Pm)   cold and humid
- tropical maritime (Tm)   warm and humid
- equatorial maritime (Em)   hot and humid

A point to be taken into account is that equatorial air is always maritime (i.e. hot and humid) and never continental (dry) in nature. Antarctic continental air (AAc) is present throughout the year over Antarctica, but the corresponding arctic continental air (Ac) exists over the Arctic only during winter.

When an air mass moves away from its source region, initially its temperature and humidity remain unchanged, but gradually they may become altered by the characteristics of the surface over which it moves. The lowermost layers will become more humid through evaporation if it passes over the sea, but dry air will remain largely unchanged if it has a long track over land or a continent. The combination of temperature and humidity also affects the stability of the air mass, which depends on its source. Polar (and arctic) air is originally stable, because it is cooled from below, whereas tropical (and equatorial) air is unstable, because it gains heat from the surface, which tends to create convection and mixing. This secondary characteristic of stability may also alter as the air moves over different areas of the surface. If tropical air moves over a colder surface, the lowest layer will be cooled and become more stable. Such cooling tends to be confined to the very lowermost layer. In contrast, arctic or polar air that moves over a warmer surface (such as the sea) will be heated and become unstable. In this case, because of the effects of convection, any warming is spread through a much deeper layer of air. Arctic continental air, for example, that moves across the sea, takes on the characteristics of arctic maritime air. (Stability and instability in relation to weather systems and the formation of clouds are dealt with in more detail in Chapter 4.)

Fronts

The boundary between two air masses with differing temperatures and humidities is known as a front. There are three different forms: cold, warm, and occluded. The last will be described in Chapter 5, when we discuss specific weather systems, but the other two are relevant here. The exact type of front is determined by which of the two air masses (warm or cold) is advancing, and

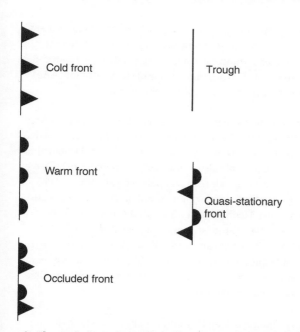

16. The symbols used to indicate the principal types of front on surface charts, together with that used to show the location of a trough of low pressure.

on weather charts is distinguished by the use of different symbols (Figure 16), indicating the location of any front at the surface. Although appearing as single lines on charts, a front is more accurately described as a 'frontal zone', a region of transition from one air mass to another with some mixing between the two. Depending on the exact nature of the air masses and individual weather systems the width of such zones may range from 10 to 200 km. The frontal systems associated with depressions are described in detail in Chapter 5.

Although there is a tendency to regard fronts as boundaries between warm and cold air (because this is the form most commonly encountered), fronts occur wherever two air masses have distinctly different temperatures. There is, for example, a semi-permanent Antarctic Front, lying at latitudes 60–65° S, between the frigid air from the interior of the continent (AAc) and the polar maritime air (Pm) further north.

In the northern hemisphere, in wintertime there is often an Arctic Front, which stretches from Greenland to northern Scandinavia, dividing frigid arctic maritime air (Am) from cold polar maritime air (Pm). A similar front (also known as the Arctic Front) between frigid arctic continental air (Ac) and polar continental air (Pc) often occurs over Canada in winter. Specific, regional fronts also occur, such as the Mediterranean front that sometimes forms between a subsidiary air mass, consisting of warmed polar maritime (Pm) air or even arctic maritime (Am) air to the north, and very hot tropical continental (Tc) air from the Saharan region. The temperature difference across this particular front may (in winter) occasionally reach 15 deg. C.

A somewhat different boundary may separate air with two different humidities. Such a 'dryline' commonly occurs over central North America, separating humid air from the Gulf of Mexico from drier air from the deserts of the south-western states, and where it separates from the warm front ahead of the advancing air from the Gulf. This dryline is an important factor in generating severe weather (such as supercells and tornadoes—see Chapter 6) over the Great Plains. A similar situation prevails in northern India (and elsewhere in the world) and again tends to create severe weather in those regions.

The Polar Fronts

The two Polar Fronts, which we have already encountered in our discussion of the global circulation, are extremely important atmospheric features. The numerous and extremely changeable weather systems that affect the Earth's temperate zones originate here, at the boundary of cold polar air and warm air from the subtropics. However, the Polar Fronts are by no means straight, but follow the Rossby waves described in Chapter 2 and encircle the globe in a series of lobes, where, at any given latitude, warm air has advanced towards the poles, and cold air towards the equator (Figure 17). The lobes consist of a series of warm and cold fronts, where the warm and cold air masses, respectively, are advancing, occasionally accompanied by quasi-stationary sections. They are particularly pronounced when the polar vortex is weak (Figure 13, bottom).

There are continual changes along the Polar Front as the lobes expand and contract. They also slowly migrate eastwards around

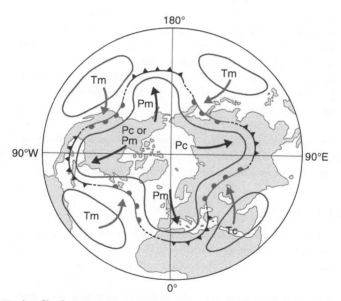

17. A stylized representation of the lobes around the north polar vortex with the alternating ridges and troughs, warm and cold fronts, and air-mass source regions. The continuous black line represents the high-speed core (the jet stream) of the Rossby waves.

the Earth, although occasionally one or more may become 'blocked', remaining stationary over a particular area for an extended period of time. Very occasionally, they may move westwards. It is rare for the boundary to remain stable for very long, and small fluctuations in position (known as secondary waves) repeatedly arise and migrate eastwards within the main flow. These secondary waves often grow into depressions (low-pressure systems, known technically as extratropical cyclones). It is these systems and the associated anticyclones (high-pressure areas) that are responsible for the extremely variable weather that is encountered over the Earth's temperate zones.

It may be noted that in North American usage it is common for depressions (extratropical cyclones) to be referred to as 'storms' and not called 'depressions' or 'lows'. Other parts of the English-speaking world tend to reserve the term 'storm' either for smaller-scale violent events (such as thunderstorms, or wind storms), or for specific destructive events (such as the 1987 October storm that affected much of southern England). Storm force 10 is also a specific range on the Beaufort scale of wind speeds (see Appendix A).

In 2015, the UK Met Office and the Irish meteorological service (Met Éireann) decided, in the 'Name Our Storms' project, to give specific names to expected severe weather events.

Jet streams

Jet streams are narrow currents of high-speed winds in the upper troposphere or stratosphere that occur where there are particularly abrupt temperature contrasts. They may be thousands of kilometres long, hundreds of kilometres wide, and a few kilometres deep. Depending on circumstances, any wind at an appropriate level that exceeds a velocity of about 25–30 ms^{-1} (90–108 kmh^{-1}) is considered to be a jet stream. (The apparent maximum speed ever recorded was 656 kmh^{-1}

above South Uist in the Outer Hebrides on 13 December 1967, although there is considerable doubt about the accuracy of this measurement.) The location of some jet streams is shown in Figure 8, but the two that have the greatest influence on surface weather are the jet streams (one in each hemisphere) at the Polar Front, where cold polar and warm tropical air masses meet. They lie at an altitude of 9–12 km (30,000–39,000 ft) and they are located on the subtropical side of the Polar Front, where there is a break in the tropopause. The locations of the Polar and the Subtropical jet streams in the northern hemisphere are shown in Figure 18.

The Polar jets

As with the general westerly flow, the jet streams weave their way around the world but, by their very nature, they are discontinuous and, as may be expected, their velocity varies along their length.

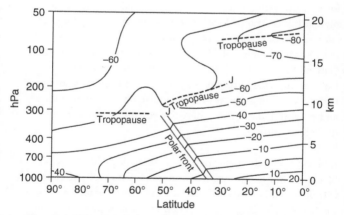

18. **The general location of the westerly Polar Front and subtropical jet streams (J) in the northern hemisphere. Typical temperatures are shown at varying heights in the atmosphere. Note the jump in temperature from north to south across the Polar Front.**

For complex reasons beyond the scope of this book, they create convergence and divergence of air at altitude as their curvature and velocity change. We have noted, for example, that there is convergence at the surface and divergence at altitude in a depression, and jet streams may act to increase or decrease these effects. They may strengthen a depression (increasing the pressure gradient at the surface, giving rise to closer isobars and stronger winds), or have the opposite effect. They may also act on the paths of depressions, as what is termed the 'steering flow', guiding them to higher or lower latitudes.

As already noted, the westerlies and the Polar Front jet streams are normally a zonal flow (known technically as showing a high zonal index), but occasionally (particularly in the northern hemisphere, and during the winter, when the winds are stronger) they exhibit major excursions in the meridional direction. These deviations frequently begin in the east and develop westwards. They may become so extreme that they lead to fragmentation of the flow (low zonal index), producing deep cold, occluding lows at low latitudes and deep warm, blocking anticyclones at high latitudes, trapped within the meanders of the jet stream (Figure 19). Occasionally, the jet stream may retreat towards the pole, leaving isolated blocking regions of high or low pressure ('cut-off highs or lows') that remain in place. Such cut-off features may lead to very persistent weather over the underlying region. A cut-off low may produce unusually cool temperatures and exceptional rainfall, whereas a cut-off high may bring drought conditions.

A blocking situation may persist for some time. In winter, a blocking anticyclone over Scandinavia, for example, often brings extremely cold arctic air south and west over Western Europe and the British Isles (Figure 20). When such a blocking anticyclone occurs, it tends to steer any depressions that approach from the west to higher latitudes than normal.

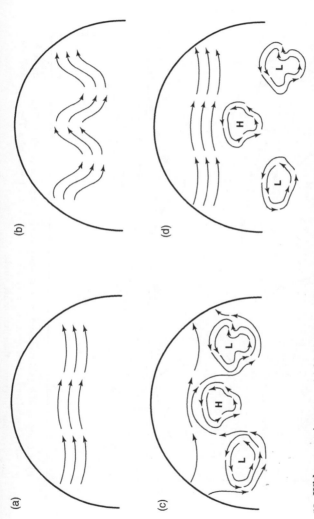

(a)

(b)

(c)

(d)

19. With a strong jet stream (a) frigid air is confined to the polar region. When the jet stream weakens, lobes of cold air move towards the equator (b). The lobes may become extreme (c) producing pockets of warm and cold air well north or south of their normal location. The jet stream may withdraw towards the pole (d), leaving behind 'cut-off' low- and 'cut-off' high-pressure systems.

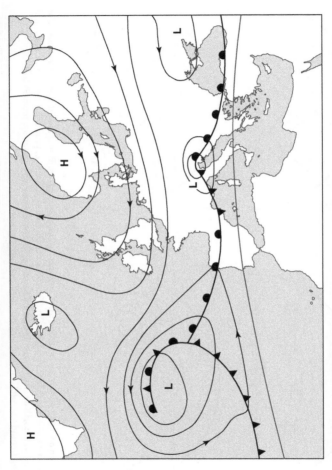

20. A typical blocking anticyclone over Scandinavia during the northern winter. This creates a persistent flow of very cold arctic air over Western Europe and the British Isles. The arrows indicate the geostrophic flow rather than the surface winds.

The Subtropical jets

The Subtropical jets are somewhat weaker than the Polar jets and lie at higher altitudes (10–16 km—32,000–52,000 ft) at latitudes of approximately 25° N and S, where there is an even greater break in the altitude of the tropopause (see Figure 18).

Other jet streams

There are other jet streams. There is an easterly tropical jet (known as the Equatorial Jet Stream) that occasionally forms in the eastern hemisphere over Asia and in the northern-hemisphere summer. It runs approximately at latitude 10° N and at an altitude of 15–20 km (50,000–65,000 ft). Here there are extreme temperature contrasts with the coldest air lying over the equatorial region, where the tropopause is highest. This jet does not extend into the western hemisphere.

There is also another broad, weak jet (the easterly African Jet) that forms in the northern summer at a similar latitude above West Africa that is much lower, 4–5 km (13,000—16,000 ft). This latter jet is significant in the development of the south-westerly monsoon winds over Africa (see Chapter 6) and also influences the creation of what are known as tropical waves, which are the initial stage in the process that may lead to systems that become tropical cyclones (hurricanes) over the North Atlantic (see Chapter 7).

Other jets may also form at great altitudes. There are, for example, the easterly Polar Night Jets, which encircle and form the equatorward boundaries of the polar vortices (described in Chapter 2). They form during the winter in the stratosphere in the winter hemisphere, and circle the Earth at latitudes of about 60° N or S, at an altitude of approximately 25 km (80,000 ft).

Chapter 4
Water in the atmosphere

Among the planets in the Solar System, Earth is unique in possessing large quantities of water, and water's properties are highly significant in determining weather. The principal reason that water has such an extremely important part to play in many atmospheric processes is because it may readily exist in three different phases (ice, liquid water, and water vapour) at temperatures that are frequently encountered on Earth. Although ice and liquid water are very familiar to everyone, water vapour is an invisible gas. Probably everyone has noticed that there appears to be a clear gap between the spout of a boiling kettle and the white steam. The 'gap' actually consists of invisible water vapour. Only where it condenses into tiny water droplets does the white steam become visible. The way in which water exists in the atmosphere is described in Box 4.

Water vapour and liquid water have an important property in that they possess latent heat. This property is most easily understood by considering a quantity of ice. Heat is required to melt the ice into liquid water, and more is needed to evaporate the liquid water into water vapour. In the reverse processes, when water vapour condenses into liquid water, and liquid water freezes into ice, heat—latent heat—is released. These transitions (in both directions) are known as phase changes, and the accompanying absorption and release of energy in the form of heat are significant

Box 4 Air doesn't 'hold' water

There is a common misunderstanding that air 'holds' water vapour, as if it were a sponge. In fact, a specific volume of gas—one cubic metre, say—at a particular temperature and pressure, contains a fixed number of molecules, regardless of the mixture of gases present. (This is known as Avogadro's Law, after the Italian physicist Amedeo Avogadro (1776–1856), who discovered it in the early 1800s.) If the air is completely dry and we ignore trace gases, just nitrogen (N_2) and oxygen (O_2) molecules are present. For every molecule of water vapour (H_2O) that is introduced in any given volume, one molecule of either nitrogen or oxygen is forced to leave. This leads to the initially apparently paradoxical situation, which many people have difficulty in believing, that humid air weighs less than dry air.

If we make the simplifying assumption that every atom (and molecule) of a particular gas has an identical atomic weight, one atom of hydrogen has an atomic weight of 1, nitrogen of 14, and oxygen of 16. The atomic weight of a molecule of nitrogen, N_2, is 28 (2×14) and one of oxygen, O_2, is 32 (2×16). Either molecule has a greater atomic weight than that of a molecule of water, H_2O, where the atomic weight is 18 ($2 \times 1 + 16$). So whether a molecule of water displaces a molecule of nitrogen or one of oxygen, humid air always weighs less than dry air. Because air consists of about 78 per cent nitrogen and 21 per cent oxygen (i.e. a ratio of approximately 4:1), five molecules of water (total atomic weight of 90) will generally displace four molecules of nitrogen and one of oxygen (total atomic weight of $112 + 32 = 144$). If two parcels of air that are in contact have the same temperature but different humidities, the humid parcel has a lower density and will tend to rise above the dry parcel.

An example of how this occurs in the atmosphere may be found when hot, dry air from the Sahara flows northwards in the wind known as the sirroco. As the air crosses the Mediterranean and encounters humid Mediterranean air, the hot wind initially hugs the surface, rather than rising above what is, in this case, the cooler air.

in many atmospheric processes, such as the formation of clouds. A way of protecting plants from the effects of a mild frost is by spraying them with a fine mist of water. When the water droplets freeze, they release latent heat and prevent the plants themselves from freezing. Ice may turn directly into vapour in the process known as sublimation. An ideal way of drying clothes is to hang them in sunlight when air temperatures are below freezing. The water freezes, but then turns directly into water vapour, leaving the clothes completely dry. Conversely, vapour may undergo a direct phase change into ice, without passing through the liquid stage. Perhaps confusingly, this vapour-to-solid change is also known as sublimation or, sometimes, as deposition. The direct change from water vapour into ice frequently occurs when frost is deposited on objects on the ground, and is largely involved in the growth of ice crystals high in the atmosphere.

Humidity and saturation

If liquid water is freely available, the number of molecules of water vapour in the air (the humidity, see Box 5) is determined

Box 5 Humidity and the mixing ratio

The terms associated with humidity often cause confusion.

Specific humidity is the ratio of the mass of vapour to the total mass of the sample, air plus water vapour.

The *mass mixing ratio* is the mass of water vapour to the mass of dry air in which it occurs. Although technically a ratio, it is often expressed as grams of vapour per kilogram of air.

Relative humidity is the ratio of actual water vapour to the amount that would give saturation at that particular temperature. It is normally specified as a percentage. It is the term that is most commonly used in forecasts for the general public, because it gives an indication of the level of discomfort likely to be experienced.

solely by the temperature. All solid and liquid substances lose vapour from their surfaces in the process known as evaporation. This vapour consists of the atoms or molecules that have sufficiently high energies to escape from the surface of the substance involved. Consider a closed container that initially holds dry air and liquid water. The velocity of the water molecules in the liquid solely depends on the temperature: the higher the temperature, the faster the molecules move, and more are able to evaporate into the air, until there is an equilibrium between the number of molecules entering the air, and the number leaving it. The humidity is then 100 per cent, and the air is said to be saturated. On cooling, more molecules condense onto the water than evaporate from it, until equilibrium is again reached. If the container were flexible, the volume of air would expand and contract on heating and cooling as discussed in Chapter 2. There is a similar relationship between water vapour and ice.

If a parcel of air containing water vapour is cooled it will eventually reach a temperature at which the vapour will condense into droplets. This temperature is known as the dewpoint, and if the parcel is in contact with the surface, the droplets will be deposited as dew. Away from the surface, droplets form only in the presence of tiny particles, known as condensation nuclei, but these are so numerous throughout the atmosphere that condensation readily takes place in the form of clouds or fog. This is not the case with freezing, where only particles of a specific shape act as freezing nuclei. When such particles are absent—as often occurs at height in the atmosphere—water vapour and cloud droplets may exist at temperatures far below 0 °C, when they are said to be supercooled. They will freeze spontaneously only when their temperature drops to –40 °C.

Frequently air is not fully saturated, but will become so on cooling. This may occur when air loses heat in contact with a cold surface or is forced to rise, when (as we have seen) the reduction in pressure causes it to expand and cool. The first process occurs when a layer of air in contact with the ground cools at night, reaches the dewpoint, and the water vapour condenses as mist or fog. Air may be forced to rise in any of four different ways:

- by convection: when heating of the ground by sunlight causes parcels of warm air (thermals) to rise
- by forced ascent (orographic lifting): when the wind forces air to rise over high ground
- by frontal lifting: when warm air is undercut by colder air, lifting it away from the surface
- by convergence: when air enters a restricted area (such as the low-pressure centre of a depression) from various directions and, being unable to accumulate, is forced to rise.

Lapse rates, stability, and instability

Although, as we have seen, there is an overall decrease in temperature from the surface to the tropopause, there may be considerable deviations from a smooth decline with altitude. The actual change in temperature with height, known as the environmental lapse rate (ELR), may be measured by a radiosonde—a balloon carrying an instrument package. Such balloons, released at specific times from stations worldwide, reveal the temperature and humidity profile of the atmosphere.

A parcel of dry (unsaturated) air that is warmer than its surroundings will rise, expand, and cool at what is known as the dry adiabatic lapse rate (DALR). ('Dry' means that condensation has not set in, and 'adiabatic' means that the parcel of air—which is a poor conductor—does not exchange heat with its surroundings.) This dry adiabatic lapse rate is greater than the typical, overall lapse rate in the troposphere, mentioned in Chapter 1, of about 6.5 deg. C km^{-1}, and equals 9.767 deg. C km^{-1}. How far a parcel of air rises depends entirely on the temperature of the surrounding air. If the environmental lapse rate (ELR) is less than the dry adiabatic rate (ELR < DALR) the rising parcel cools more quickly than its surroundings. When the temperatures become equal, the parcel's buoyancy ceases and it comes to a halt. If the parcel were forced to rise still further (forced by the wind over a mountain range, for example) it would become colder than the surrounding air and (provided it had not become saturated in the process) would tend to sink back to the level it had before it encountered the obstacle. Such conditions are said to be stable and, if the parcel has reached the dewpoint at that level, would lead to the formation of clouds. In the absence of convection, such clouds would be layer clouds (stratiform), with a base at the condensation level and a limited depth. Clouds of this sort (often the type known as stratus) frequently shroud the tops of hills, mountains, and isolated islands.

If, on the other hand—and partly dependent on the relative humidity—the environmental lapse rate is greater than the dry adiabatic rate (ELR > DALR), the parcel of air, despite expanding and cooling, remains warmer than its surroundings and continues to rise. Such conditions are unstable and lead to the formation of cumuliform clouds, particularly cumulus and cumulonimbus. The rate of ascent may even accelerate, especially if the parcel reaches the dewpoint and condensation occurs. The release of latent heat in the parcel increases the temperature difference between it and the surroundings, causing it to rise even faster. The parcel now cools at the saturated adiabatic lapse rate (SALR) which, depending on the exact conditions, lies between 4 and 7 deg. C km^{-1}. (At higher temperatures, more water vapour is available to condense, releasing more latent heat, so the numerical value of the SALR is less, and the decrease in temperature with ascent becomes slower. The lower the temperature, the less water vapour and latent heat is available.)

If precipitation occurs in the form of rain, snow, or ice crystals that fall out of the cloud, heat is lost in the precipitated material. The parcel of air may still continue to rise, but at a 'pseudo-adiabatic' rate, which is slightly greater than the SALR.

As mentioned in Chapter 3, the thin layer closest to the Earth's surface may experience extreme heating from the ground, giving rise to a lapse rate that is far greater than the dry adiabatic lapse rate. Such an elevated rate is known as a 'superadiabatic lapse rate'.

Descending air warms at the appropriate rate. If air has been forced to rise over hills, but no condensation has occurred, then it warms at the dry adiabatic rate on descent. Similarly, clouds that form over the tops of hills, but give no precipitation, will evaporate as the air descends on the leeward side, initially warming at the saturated rate. If precipitation has occurred, removing water (and heat) from the parcel of air, warming at the dry adiabatic rate will begin sooner than would otherwise be the case, and at any given level, the air may be much warmer than it was on its ascent. This is one cause of föhn conditions, which are described in Chapter 8.

It is frequently the case that precipitation in the form of rain or snow falls on the windward side of a range of mountains or a line of hills, so that much of the moisture is deposited there, and little precipitation falls on the leeward side. This creates a 'rain shadow', which is very commonly found downwind of mountains or hills. Where the prevailing winds are westerly—as occurs over most of the temperate zone—conditions to the east of mountain ranges (or even quite moderately high hills) are usually much drier.

The origins of rain

As a parcel of air continues its ascent, it will eventually reach a level at which its temperature falls below freezing, in the process known as glaciation. As mentioned earlier, condensation nuclei are found in vast numbers throughout the atmosphere, but freezing nuclei are far less common, because they must be a specific shape for water molecules to freeze onto them. Because freezing nuclei are relatively rare, cloud droplets are often supercooled, and most ice crystals (all hexagonal in form) originate at temperatures slightly below 0 °C. Different types of crystals arise within very

Table 2 Ice crystal forms

Form	Temperature range	Shape
Thin flat plates	0 °C to −4 °C	
Needles	−4 °C to −6 °C	
Hollow columns	−6 °C to −10 °C	
Sector plates	−10 °C to −12 °C	
Dendritic plates	−12 °C to −16 °C	
Sector plates	−16 °C to −22 °C	
Hollow columns	below −22°C	

specific ranges of temperatures, but the most efficient freezing nuclei promote freezing between −10 °C and −15 °C. Many different forms of ice crystal arise at different temperature ranges and these are shown in Table 2.

Dendritic plates are the type most commonly called 'snowflakes', the design seen everywhere at Christmas. Sector plates consist of six unbranched, flat plates, joined together in a star shape. Only when conditions are very cold do any of these forms reach the ground intact. The flakes of snow that normally fall may be several centimetres across and actually consist of many individual crystals that have frozen together.

Cloud droplets are so small (generally about 20 micrometres, 20 μm, in diameter) that they tend to remain suspended in the air and rarely grow through collisions with other cloud droplets.

The smallest raindrops are about 2 mm in diameter, so some one million cloud droplets are required to create even the smallest raindrop. The precise way in which rain originated was thus a puzzle, until the Swedish meteorologist Tor Bergeron proposed glaciation as the primary cause. His theory was later extended by W. Findeisen, and the theory is commonly called the Bergeron process, the Bergeron–Findeisen process, or even the Wegener–Bergeron–Findeisen process, after the contribution of Alfred Wegener, the meteorologist, most famous (in geology) for originating the concept of continental drift.

Most rain that falls in temperate and polar regions originates as ice crystals that have formed by glaciation in the upper regions of clouds. Ice crystals grow at the expense of supercooled water droplets and eventually become heavy enough to fall. They may then grow through collision with more supercooled water droplets, which freeze instantly on contact. The crystals eventually melt into raindrops as they fall into warmer layers of air and may continue to grow through collision with other water droplets. As drops fall they tend to assume a flattened, 'bun-like' shape and fragment into smaller droplets if they become too large. The largest raindrops ever recorded were about 10 mm across, but most are much smaller.

For many years, the Bergeron glaciation process was considered to explain all forms of rain, until it was realized that many tropical clouds, the source of heavy rain, did not reach the freezing level. It was found that under conditions of extremely vigorous convection, collisions and coalescence of cloud particles did occur leading to the growth of raindrops from liquid water. The coalescence process is now known to occur in some temperate clouds (cumulonimbus and deep cumulus congestus) during summer.

The two processes, glaciation and coalescence, are sometimes referred to—quite seriously—as the 'cold-rain' and 'warm-rain' processes.

Clouds

There are ten main types (genera) of clouds, and these may be defined both by means of their altitude, and also by their structure. Three ranges of height (known technically as étages) are officially recognized: low clouds with bases from ground level to 6500 ft (approximately 2000 m or 2 km); medium clouds with bases between 6500 ft and 20,000 ft (approximately 6000 m or 6 km); and high clouds with bases at 20,000 ft or above. Cloudbase is, of course, the height at which a parcel of air reaches the dewpoint and condensation occurs.

Table 3 shows the ten cloud types and the heights at which they occur, although it should be noted that cloudbase tends to be lower at high latitudes and in winter. Brief details of the different cloud types are given in Appendix B.

Although nimbostratus, the main rain-bearing cloud in depressions, is classed as a medium-level cloud, it may extend downwards, essentially as far as the surface. Cumulonimbus

Table 3 Cloud types

HIGH CLOUDS (bases 20,000 ft or above)		
Cirrus	Cirrocumulus	Cirrostratus
MEDIUM CLOUDS (bases between 6500 ft and 20,000 ft)		
Altocumulus	Altostratus	Nimbostratus
LOW CLOUDS (bases below 6000 ft)		
Cumulus	Stratus	Stratocumulus
CLOUDS THAT EXTEND THROUGH MORE THAN ONE LEVEL		
Cumulonimbus		

21. Cumulonimbus incus. Two cumulonimbus clouds that have reached to inversion at the tropopause (actually a low winter tropopause) and spread out into the characteristic 'anvil' shape. (Known as cumulonimbus 'incus' after the Latin word for 'anvil'.)

clouds (shower clouds) may have low bases, but often extend upwards as far as the tropopause at 60,000 ft or more. In many cases, upward growth is arrested when the cloud towers reach the inversion at the tropopause, and the clouds then tend to spread out into a characteristic anvil shape (Figure 21). Particularly vigorous cumulonimbus cells may possess sufficient energy to penetrate the tropopause, producing domes of cloud—'overshooting tops'—in the lowermost stratosphere.

Although details need not concern us here, in a way similar to the classification of animals and plants, each cloud type (with one exception: nimbostratus) is subdivided into species and varieties based on the structure and general nature.

Overall, clouds are arranged into two main groups: cumuliform, or heaped clouds, and stratiform, or layer clouds. Cumuliform clouds arise through convection and include cumulus, cumulonimbus, stratocumulus, altocumulus, and cirrocumulus. Stratiform clouds generally arise through lifting of a humid layer, and include stratus, nimbostratus, altostratus, and cirrostratus. (A third group, cirriform clouds, is sometimes recognized. It includes the ice-crystal clouds cirrus, cirrocumulus, and cirrostratus.)

Stratocumulus, altocumulus, and cirrocumulus clouds occur in layers, but are broken up by weak convection into individual cloudlets in the form of clumps, rolls, or plates of cloud. They may therefore be regarded as exhibiting both stratiform and cumuliform characteristics.

Stratocumulus, altocumulus, and cirrocumulus cloud layers often arise when rising air encounters a temperature inversion, preventing further upward movement. The air then tends to spread out beneath the inversion and, if it has reached the dewpoint, creates a layer of cloud. If convection continues, the individual cloudlets may spread out and merge to give a continuous layer of cloud, becoming stratus, altostratus, or cirrostratus.

The formation of clouds

Cumuliform clouds (particularly cumulus and cumulonimbus) develop through convection in the form of warm thermals that rise from heated ground. Similarly, stratocumulus, altocumulus, and cirrocumulus often arise when thermals reach an inversion or stable layer and the rising air spreads out sideways into individual cloudlets, separated by clear air around the edges where the circulation causes the air to descend. But these three types also develop from stratiform clouds, when the top of the layer radiates heat away to space. The cooled air sinks, breaking up the layer into separate sections. An identical mechanism of 'upside-down convection' produces mamma, deep pendulous pouches of cloud often observed beneath the overhanging anvils of cumulonimbus clouds.

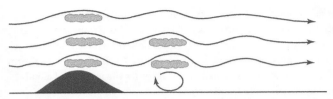

22. When air passes over a range of hills it creates waves downwind. If the air is humid, wave (lenticular) clouds may arise in the crests of the waves. Sometimes a rotor may be created beneath the highest wave, where the surface wind direction is reversed.

Cloud may, obviously, arise when the gradient wind forces parcels of humid air to rise over high ground, reaching the condensation level as it does so. The subsequent behaviour of the cloud entirely depends on whether the air is stable or unstable at that level. With stable air, the cloud will often dissipate as the air sinks to leeward of the obstruction. Frequently, a train of waves of rising and sinking air will be created downwind, and cloud will often form in the crests of the waves (Figure 22). Such wave-trains and wave clouds may extend for many kilometres downwind of the high ground (Figure 23) and the series of clouds are frequently visible in satellite images.

When the air is unstable it may continue to rise after the initial impetus given by the high ground. Cumulus or cumulonimbus clouds may then build up over the peaks. Cumulonimbus that have built up over high ground, in particular, may give rise to major downpours and have, in some cases, created devastating flash floods. Particularly devastating flash floods arising in this way in the United States have arrived so suddenly that they have caused numerous deaths. Another example, luckily with no casualties, occurred when cumulonimbus clouds that had grown over the high ground in the centre of the Cornish peninsula were the source of the extreme precipitation that resulted in the floods that swept through Boscastle in 2004.

23. **Altocumulus lenticularis.** Lenticular altocumulus clouds forming in the tops of waves created by distant hills (off to the left of the image). Suggestions of a similar, even higher, cloud may be seen, together with lower cumulus and stratocumulus cloud.

Cloud streets

Satellite images (Figure 24) often show a striking pattern of clouds, known as cloud streets (or, more formally, as horizontal convective rolls or horizontal roll vortices), of more or less parallel rows of cumulus or cumulus-type clouds. Cloud streets are particularly noticeable in satellite images where air is streaming off the land and over the sea. Such lines of cloud are also visible (albeit less clearly) from the ground, where they also occur over land, and may often be linked to specific sources of thermals that produce a succession of clouds that are carried downwind. The process leading to the formation of the highly regular pattern of rolls is complex and still poorly understood. The rolls form within the planetary boundary layer, where this is capped by an inversion. The air within adjacent rolls of cloud is rotating in opposite directions. Cloud arises where the air is rising, and disperses, to give the clear spaces, where the air is descending.

24. Cloud streets forming over the Black Sea when cold air blows over warm water. NASA image from Aqua satellite.

Chapter 5
Weather systems

At temperate latitudes, such as those of the British Isles, the most significant changes in the weather, with major changes in wind strength and direction, as well as rainfall, are associated with the passage of depressions (low-pressure systems), more formally known as extratropical cyclones. The exact changes that occur will, of course, depend upon the actual track of the low-pressure centre. As mentioned in Chapter 3, the path of the depression and whether it strengthens or weakens may depend on the exact course of any jet stream above it, as well as the location of any neighbouring high-pressure systems, particularly blocking anticyclones.

Changes in cloud cover, wind direction, and rainfall are generally regarded as being associated with advancing warm or cold fronts, but, perhaps unexpectedly, there may be significant clouds along a stationary front. Even though the front itself may not be moving, the boundary between the two air masses is rarely vertical and warm air normally overlies the edge of the cold air mass, which may cool the warmer air sufficiently for the temperature to drop to the dewpoint and clouds to form. There may also be local mixing of the two air masses, which may contribute to the formation of clouds.

When it comes to active depressions, however, although there may be significant precipitation (rain or snow) whether the pressure centre passes north or south of the observer's location (or over it), the strongest winds tend to be found on the southern flank of the centre (in the northern hemisphere). The severity of the rainfall and wind strength will also depend on the stage that the depression has reached in its evolution.

The development of depressions

An idealized representation of the way in which depressions (low-pressure centres) develop in the northern hemisphere is shown in Figure 25. Because the situation on a quasi-stationary front is not stable in the long term, small waves tend to develop. Such an initial instability (a secondary wave) on the front (Figure 25a) gives rise to distinct warm and cold fronts (Figure 25b). A closed circulation develops around a low-pressure centre and a distinct, triangular feature, known as the warm sector (Figure 25c), develops between the advancing warm front, where air is rising above the cooler air to the north and east, and the following cold front to the north and west, where dense, cold air undercuts the warm air. The cold front advances faster than the warm front and eventually reaches it (Figure 25d), lifting it away from the surface and giving rise to an occluded front, where there are three distinct air masses, with different temperatures. There is a pool of warm air aloft. The point where the three fronts meet is known as the triple point.

The term 'triple point' is used in three different contexts in meteorology. One is as here, where it indicates where three frontal systems meet. Another, very important in discussion of the behaviour of water in the atmosphere, is the unique temperature at which all three phases (solid, liquid, and gaseous) of a substance exist simultaneously. For water, the triple point is defined scientifically as 273.16 K (0.01 °C), fractionally higher than the freezing and melting point 273.15 K (0 °C). The final use of the term occurs in tropical meteorology, where it describes the point at which three distinct air masses are in contact. Such

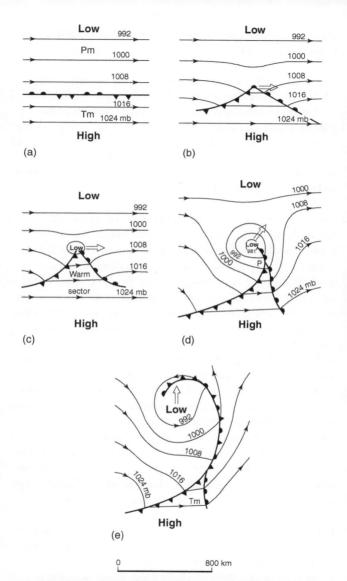

25. **An idealized representation of the evolution of a low-pressure system (a depression or 'low'). For details of the different stages see the text.**

locations favour the formation of tropical cyclones (the latter being discussed in Chapter 7).

Finally, the system starts to decay (Figure 25e) as virtually the whole of the warm sector is undercut. The depression 'fills' and the central pressure increases. Generally, the Polar Front has pushed further south behind the depression and the next low may form at a lower latitude. Eventually, the succession breaks off, and the Polar Front is re-established at a higher latitude, at which stage the overall process repeats. A satellite image of an occluded and decaying system may be seen in Figure 26.

The whole depression usually travels eastwards and slightly to the north, carried by the general westerly flow, and normally approximately parallel to the isobars in the warm sector. If a strong, secondary low has formed on the trailing cold front, however, the two low-pressure centres may tend to swing round one another, with the secondary progressing eastwards, while the centre of the older, decaying low may shift back towards the west. In many cases a whole family of depressions occurs, following one another towards the east (Figure 27). Such depressions form to the south of the jet stream (in the northern hemisphere, of course), and the jet itself develops an 'S' shape, overlying the centre of each depression. The high cirrus clouds in the jet stream often give advance warning of an approaching depression because, well ahead of the warm front, they may be observed racing towards the east or south-east. They often lie more or less parallel to the warm front, but are 'crossed' with respect to the wind at the surface (which might be southerly or south-westerly), sometimes even moving at right angles to it. This 'crossed-wind' factor is yet another indication of an approaching depression system.

In the earliest stages (Figure 25b), as a wave starts to develop on the Polar Front and air begins to flow across the isobars, winds are light. In the classic form of frontal system, the air in the warm sector is rising at both fronts (known technically as 'anabatic

26. A Meteosat image of a depression over Western Europe on 28 March 2016. The mass of cloud over Italy and the Carpathians lies on the warm front, and the convective cloud behind the cold front is clearly visible, as is the occluded front leading to the depression centre.

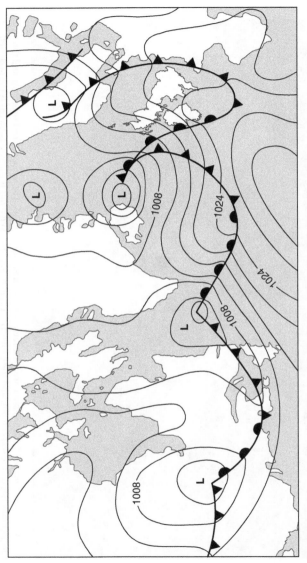

27. A family of depressions at various stages of development as they form on the Polar Front.

fronts' or 'anafronts'. The situation may arise where the warm air is subsiding, giving rise to weaker systems (with 'katabatic fronts' or 'katafronts'), which are described later. The terms 'anabatic' and 'katabatic' and compounds (both derived from Greek words) are generally applied by meteorologists to the motion of air upwards and downwards, respectively. Katabatic winds, for example, are downslope winds, perhaps flowing off an ice field, and some of which are also known as 'mountain winds'.

Very few fronts are as clear-cut as any description may suggest and the majority display a complex mixture of features with segments where the warm air may be rising and others where it may actually be sinking. In what follows, it is assumed that the fronts are reasonably uniform and exhibit the 'classic' structure.

The depression sequence

Although fronts are drawn on charts as a single line, in reality they consist of a complex three-dimensional feature, with variations in the location of the sloping frontal zone, where there is a transition from one air mass to another. (This frontal zone is variable in depth but is usually at least 100 km wide at the surface). The front typically advances faster over some areas of the surface, so there is generally an irregular boundary between the two air masses. At any front there is normally a distinct change in the direction of the isobars and as a result, when a front passes through, there is a significant change in wind direction. In the northern hemisphere, the wind usually veers at both warm and cold fronts.

The terms 'veering' and 'backing' are used to describe changes in wind direction. A change in a clockwise direction (say, from south to south-west) is known as veering, and one in an anticlockwise direction as backing.

The rising air at the warm and cold fronts leads to the formation of clouds. Initially, warm air from the southern side of the Polar Front (in the northern hemisphere) has only begun to rise above the cooler air to the east, so although clouds do form as the warm air rises, cloud cover is often broken with layers of cirrocumulus and altocumulus. Rainfall is light or absent. Any clouds in the air ahead of the depression (often scattered cumulus) tend to become flattened as warm air moves in above them, creating a temperature inversion and suppressing convection. There is usually rather greater activity at the advancing cold front and a greater likelihood of significant rain.

Later, at what is known as the open stage (Figure 25b), a distinct warm sector develops between the cooler air to the east and the cold polar air to the west.

The warm front

A typical warm front (Figure 28) has a shallow slope of between 1:100 and 1:150, whereas a cold front is much steeper at 1:50 to 1:75. The wedge of warm air at a warm front may reach the tropopause some 1000–1500 km ahead of the actual frontal zone at the surface, which (as just noted) may be between 100 and 300 km in width. The resulting high clouds provide a useful clue to the impending change in the weather. As a warm front approaches the surface pressure drops at a gradually increasing rate. The air temperature also decreases.

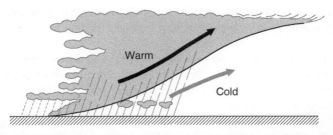

28. **An idealized warm front, showing the typical sequence of layer clouds that occur ahead of the surface front.**

As the warm front advances, the clouds become more organized and stratiform in nature, with cirrus and cirrostratus in the lead. The thin sheet of ice-crystal cloud in cirrostratus often gives rise to various optical phenomena, although these frequently go unnoticed by the general public. The most common effect consists of a halo around the Sun, with a radius of 22°, arising because of the refraction of sunlight through the regularly shaped ice crystals. The cirrus and cirrostratus thicken into altostratus, which in turn thickens, and the cloudbase lowers, giving rise to nimbostratus cloud. Altostratus is often a mixed cloud, consisting of both ice crystals and (usually supercooled) water droplets. Generally, ice crystals falling from the altostratus initiate rainfall in the lower nimbostratus cloud and significant amounts of rain may reach the ground. The nimbostratus itself becomes thicker as the front advances and its base descends towards the surface and may even reach it (particularly over high ground). Sometimes the early cloud may be cirrocumulus, gradually turning into altocumulus, which then thickens into a layer of altostratus, and finally into the rain-bearing nimbostratus.

The warm sector

Behind the warm front, within the warm sector, the cloud may be more or less broken, depending on the distance from the depression's centre. Frequently, especially over Britain, the warm air may be relatively stable, so stratiform cloud may persist, often in the form of stratocumulus. The air pressure tends to steady, unless the depression is in the process of actively deepening, and the temperature normally rises.

The cold front

The stability of the air in the warm sector also affects the cloud cover at the following cold front. With stable air, the frontal cloud may resemble a mirror-image of that at the warm front with nimbostratus and significant rainfall, followed by higher altostratus

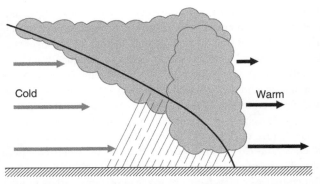

29. A typical 'passive' cold front, with a band of heavy rain, followed by layer clouds that gradually increase in height behind the front.

and cirrostratus, which may become more or less broken as the front passes. A 'classic' cold front (Figure 29) exhibits a line of deep convective (cumulonimbus) cloud with heavy rainfall, which will usually be followed by a relatively narrow belt of medium-level and higher cloud, ahead of the main cold air mass. The air pressure may decrease slightly ahead of the front, but increases suddenly as the front passes through. There is an abrupt drop in temperature with the arrival of the cold air mass behind the front.

As a depression develops, the cold air to the west eventually overtakes the warm front and completely undercuts the warm air, raising it above the surface. The cold front reaches the warm front, and a pool of warm air forms above the surface together with an occluded front that forms on its eastern side (Figure 25c). As mentioned earlier, the point where the three fronts meet is known as the 'triple point'.

Occluded fronts

The exact form of an occluded front depends on the relative temperatures of the leading and following cold air masses

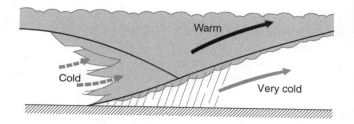

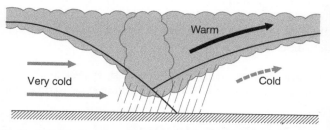

30. **Occluded fronts, where the warm air has been lifted away from the surface. In a warm occlusion, the coldest air is ahead of the front (top). In a cold occlusion, the coldest air is behind the front (bottom) and there tends to be more vigorous activity in the frontal zone.**

(Figure 30). If the air ahead of the front is coldest, a warm-front occlusion results (Figure 30 top). A cold-front occlusion is produced when the air encroaching from the west is the colder (Figure 30 bottom). There is a tendency for active cumulonimbus clouds (and the resultant pulses of very heavy rain) to be embedded within cold-front occlusions. Both forms may give rise to heavy rain. Some occluded fronts trail long distances behind the triple point, wrapping right round the low-pressure centre, sometimes producing a multi-ring spiral. Depending on the exact track of the depression they may give rise to extended periods of seemingly endless rain and often result in flooding.

Although diagrams of frontal systems show air flowing towards the fronts, it is obvious that such air cannot accumulate indefinitely.

It has been established that the main airflow in depressions occurs in what are termed 'conveyor belts'. The main flow of warm humid air, in the warm conveyor belt, begins at an altitude of about 1 km and gradually ascends, flowing approximately parallel to the cold front, but then turns to the right (in the northern hemisphere) as it passes over the warm front at an altitude of 5–6 km. There, it combines with another flow of air that originates in the middle troposphere, behind the cold front. It is the warm conveyor belt that transports essentially all the water vapour that falls as rain or snow along the warm frontal zone. There is yet a third conveyor belt of cool air that originates ahead of the warm front, turns to the right, and flows beneath the warm conveyor, before rising into the occluded front (Figure 31).

The area affected by rain (or snow) obviously changes as a depression develops. A typical rainfall pattern at the warm-sector

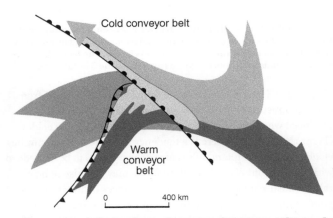

31. The conveyor belts in a depression system that transport heat and humidity. Most of the humid air is in the warm conveyor belt, which starts low ahead of the cold front, rises, then eventually turns above the warm front. Cool air from the middle troposphere behind the cold front also combines with this flow. A third conveyor belt moves cool air from ahead of the warm front, passes beneath the warm conveyor, and rises into the occluded front.

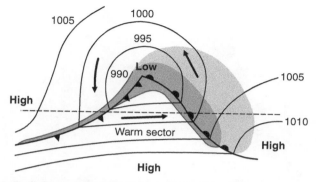

32. Typical areas affected by light and heavy rain when there is a closed circulation and a well-developed warm sector.

stage is shown in Figure 32. The persistent precipitation ahead of the warm front is often spread over a very wide area, 200 to 300 km wide. Within that region, bands of heavier and lighter rain tend to lie parallel to the front. There may sometimes be pockets of convection hidden within the overall warm-frontal structure—and thus invisible to casual observers on the ground—that give rise to pockets of intense rainfall. At the cold front there is often a band of very heavy rain some 50 km wide, frequently preceded by an area of lighter rain (Figure 33).

Katabatic fronts

As mentioned in Chapter 3, as the jet streams weave their way around the world they may cause the convergence or divergence of air in the middle troposphere. When there is convergence at altitude, the accumulation of air escapes by descent, which may cause the air within the warm sector of a depression to subside. This produces subdued katabatic fronts, where the characteristic cloud is thick stratocumulus. As a warm front approaches, for example, low clouds may gradually thicken into dense stratocumulus which, however, will only give rise to fairly

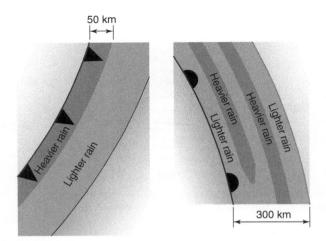

33. Typical patterns of precipitation at warm (right) and cold (left) fronts. Note the bands of heavier rain embedded within more general, lighter rain at the warm front.

light rain or drizzle. A similar type of thick stratocumulus will accompany the cold front, with little (or no) convective activity. In the cold air behind that front, however, there may be considerable convection with showers or thunderstorms.

North or south of a depression centre

If the centre of the depression passes well to the south or north of the observer, the sequence of clouds and the ensuing rainfall will exhibit a different pattern. To the north of the centre, the upper wind (the jet stream) will be westerly, whereas the surface wind will be in the opposite direction, from the east. The cloud will tend to be cirrus that gradually becomes cirrocumulus, but then begins to disperse. Pressure usually drops slightly. The surface wind tends to back, sometimes swinging round from east to north-east or north, and pressure gradually rises as the low-pressure centre passes to the south.

Well to the south of a depression centre, where the systems' motion just carries the warm sector across the observer's location, the 'normal' sequence of clouds is also modified. Cirrus may increase to cirrostratus and then altostratus or altocumulus, but rarely do the clouds cover the whole sky. Changes to pressure and wind direction are small and occur slowly. Often there is a ridge of high pressure behind the warm front, and no readily detectable following cold front.

Isolated fronts

Both warm and cold fronts may occur on their own, not associated with a low-pressure area. This is particularly the case over continental areas where large air masses may arise. Both fronts are generally similar to those already described, although warm fronts may display more convective activity. Cold fronts advancing over strongly heated land may be very active and very abrupt, showing deep convective clouds, and a narrower band of overall cloud (Figure 34). There is normally a significant drop in atmospheric pressure ahead of the front, and a sharp rise behind it. In winter, when, for example, there is an incursion of polar air from Canada over the United States, because the land is cold, there may be greatly reduced convective (shower) activity behind the advancing cold front.

Intensification of depressions

Although we have described the way in which depressions are formed, occasionally they may undergo sudden deepening, especially when affected by other major weather systems. Over the Atlantic, for example, decaying hurricanes, tracking north-eastward, may combine with an existing depression and cause dramatic deepening. When the central pressure drops by at least 24 hPa in twenty-four hours, the resulting system is described as a 'bomb'. Not unexpectedly, such bombs are usually accompanied by extremely high winds.

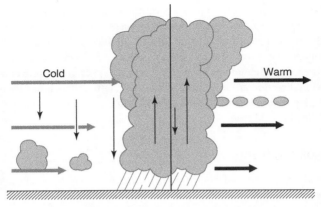

34. An 'active' cold front, where the frontal surface is roughly perpendicular to the ground, and which is accompanied by vigorous activity.

Again over the Atlantic, incipient depressions off the East Coast of the United States may be intensified by the incursion of very cold polar or arctic air from Canada, resulting in violent 'north-easters' ('nor'easters') that bring violent winds and extreme rain- or snowfall to New England and neighbouring states.

A discovery after the Great Storm of 16 October 1987 is that deep depressions may be accompanied by what is termed a 'sting jet' of extremely high winds, near the end of the curl of cloud that wraps round a depression centre. This is a narrow band—perhaps no more than 50 km across—that hits the surface with damaging winds of 150 kph or more. It begins in the middle troposphere and descends toward the surface, cooled and accelerated by the evaporation of rain or snow that falls into it.

Thermal and polar lows

Non-frontal low-pressure regions may also arise, such as when there is intense heating over land, particularly in summer. This may even lead to a low-pressure centre and a circulation with closed isobars. This is known as a thermal low or thermal depression, but frequently the heating is insufficient to create a closed circulation, and the effect is to distort the overall pattern of isobars and create what is termed a thermal trough. Both thermal lows and thermal troughs tend to decay when surface heating disappears at sunset. They may, however, be sufficiently strong to create instability with showers and thunderstorms.

A similar, but usually more intense, situation may arise when cold polar air sweeps towards the equator over relatively warm seas. This is particularly common behind the cold front of a large occluding depression (Figure 35). In this case the supply of heat from the surface does not die away at night and such polar lows (or 'polar depressions') may become very strong. As with thermal lows, weaker heating may create a low-pressure trough, but both polar lows and troughs may be the site of intense convection and produce very heavy showers or longer periods of rain or snow (Figure 36).

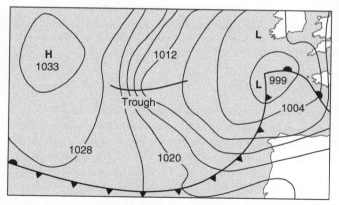

35. A trough line developing in the cold air mass behind the cold front of a depression. Such troughs may go on to become secondary depressions ('polar lows') in their own right.

36. A well-developed polar low to the south of Iceland. The low-pressure centre is surrounded by cloud and, because of the way in which the low has formed, there is no warm sector.

Atmospheric rivers

In recent years a previously unknown mechanism has been discovered that transports large quantities of water vapour across the globe. These 'atmospheric rivers' are narrow bands of highly humid air in the middle troposphere that may transport as much as 20 per cent of the total amount of water vapour moving from the tropics to higher latitudes. They appear to arise when deep depressions draw a narrow stream of air ahead of the cold front, which later becomes the stream previously described as forming the warm conveyor belt.

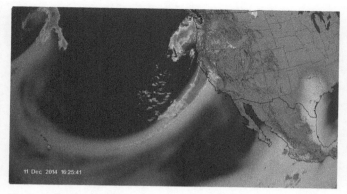

11 Dec 2014 16:25:41

37. A satellite image obtained in a channel showing water vapour of an atmospheric river crossing the coast of California on 12 December 2014. This produced large amounts of rain with heavy snowfall over the Sierra Nevada.

Although these atmospheric rivers tend to deposit most of their water over the oceans, they may reach land and, depending on the exact location and situation, may give rise to major rainfall and flooding. A high-pressure area, for example, may force the river to follow a particular route. Large quantities of precipitation may occur if a river encounters a mountain range and this is one cause of major floods in California, when the atmospheric river encounters the Sierra Nevada (Figure 37). Similar conditions are believed to have contributed to the major flooding in Cumbria in 2009, Cornwall in 2012, and possibly to the extreme flooding in Cumbria, Lancashire, Yorkshire, and elsewhere in Britain in late 2015. An atmospheric river over the Atlantic approached the eastern United States in 2010 and encountered a strong squall line. The combination deposited 300–500 mm of rain over Tennessee, and flooding caused eleven deaths in Nashville.

High-pressure systems

The weather associated with high-pressure systems is much quieter. As mentioned earlier, anticyclones may be classed as 'cold'

87

or 'warm highs', with an outflow of cold or warm air, respectively, at the surface. Cold highs are shallow accumulations of cold air that tend to form during the winter in polar regions or over continental interiors, although in other regions the high pressure tends to be in the form of a ridge, with no closed circulation. The air is usually very cold and skies are often completely clear with temperatures dropping very low at night. Sometimes cold highs are capped by an inversion that limits the growth of cloud during the day, and scattered cumulus may spread out into stratocumulus.

In warm highs, the air, descending throughout the depth of the troposphere, tends to suppress any cloud formation (Figure 38), although there may be some scattered cumulus or broken stratocumulus. When the high is underlain by a warm, humid, tropical maritime air mass, extensive areas of stratus cloud or

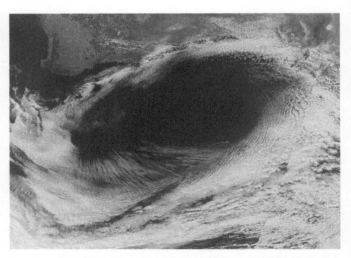

38. A satellite image of a most unusual anticyclone, situated to the south of Australia, with completely clear skies in the centre, observed on 8 September 2012. Because this is in the southern hemisphere, the circulation around the high is anticlockwise.

fog may form, especially with a fall of temperature at night. In summer, daytime heating is often enough to mix the layer of air and disperse any low cloud or fog, but in autumn and winter, in particular, the cloud or fog may be very persistent. Sometimes overcast skies with thick stratus or stratocumulus may last for days on end, giving rise to what has been called 'anticyclonic gloom'. The stagnation of air in the centre of anticyclones often results in the trapping of pollutants, and this situation may become particularly severe where the land forms a natural basin, preventing any air from escaping at the surface.

The deep column of descending air in a warm high tends to form a barrier to the normal flow of air in the westerlies and to the eastward progression of depressions. The anticyclone becomes a blocking high and exerts a great influence on the weather in surrounding areas, steering depressions to higher or lower latitudes than normal. A situation that commonly affects Western Europe in winter is when a blocking high develops over Scandinavia (Figure 20). The clockwise circulation of air around the high brings bitter easterly winds to Western Europe. Some depressions may be forced into a more northerly track, but often they are forced round to the south, bringing unusually wet and windy conditions to Iberia and the western Mediterranean.

Chapter 6
Weather in the tropics

The weather systems described in Chapter 5 are typical of the weather over the temperate zones, with dominant westerly winds and a succession of depression systems and occasional, relatively quiet periods when anticyclonic conditions occur. Particularly strong depressions and the accompanying storms are encountered during the winter and overall conditions are greatly affected by the strength of the polar vortices, with incursions of cold arctic air and blocking situations occurring when the vortex weakens, resulting in major excursions of the jet streams to lower latitudes. By contrast, weather in the tropics, here considered to be the belt between the two subtropical anticyclones, lying at approximately latitudes 30 °N and S, rather than the zone between the Tropics of Cancer and Capricorn, is characterized by rather different conditions.

The trade winds

The trade winds consist of air that flows out of the subtropical anticyclones towards the equatorial trough (the thermal trough). They are the north-east trades in the northern hemisphere, and the south-east trades in the southern. They are strongest in the winter season, tending to weaken during the summer. In addition, during the positive phase of the Arctic Oscillation (the AO), described in Chapter 9, the north-easterly trades are strongest, weakening when the AO enters a negative phase. Because the trade winds originate in regions of subsidence, the regions closer to the subtropical anticyclones and the eastern portions of the trade-wind belt tend to be drier and experience lesser cloud cover. Closer to the equator and further west over the oceans, the air takes up more moisture, giving rise to more cloud and showery conditions. Typical trade-wind clouds are cumulus, which are often confined in height by the trade-wind inversion, which, because the air above is subsiding, often occurs at the relatively low altitude of 450 to 600 m (1500–2000 ft), although some vigorous clouds break through this inversion. Any rainfall is usually initiated by the coalescence ('warm rain') process, described in Chapter 4.

The Intertropical Convergence Zone

As already mentioned, one dominant feature of the tropics is the equatorial trough of low pressure, where the two sets of trade winds converge at the Intertropical Convergence Zone (ITCZ). Here warm, moist air forms the rising limb of the Hadley Cell. However, the ITCZ is by no means a consistent feature as might appear from Figure 12, which shows the average positions in January and June. It varies considerably and is often discontinuous, with the regions of convergence expanding and shrinking in area. These individual areas may remain stationary or move westward. The changes in the strength and position of the ITCZ are particularly noticeable over the oceans. Over land, the ITCZ tends to move north and south with the seasons, whereas over the oceans its motion is affected by the sea-surface temperature. Over the Pacific, in particular, there are sometimes twin convergence zones, north and south of the equator, usually unequal in strength, with a narrow band of high pressure between the two. In the eastern Pacific, west of South America, the convergence may be considered an extension of the ITCZ over the Atlantic (and is called by that name). There is a separate, very important convergence zone, known as the South Pacific Convergence Zone (SPCZ), that extends east from the eastern end of Papua New Guinea to the middle of the Pacific at around 30 °S and longitude 120 °W. The SPCZ is highly variable and is closely associated with the changes involved in the Walker Circulation and El Niño events (both described in Chapter 9).

As suggested in Figure 12, the convergence zones over the Pacific and Indian Oceans (in particular) show major changes in location during the northern summer, and these are related to the seasonal monsoons, discussed shortly. In summer, the ITCZ over the Atlantic lies at about latitude 10 °N but in all cases normally lies some degrees of latitude equatorward of the 'thermal trough' where solar heating is at a maximum. The ITCZ is marked by intense convection and cloudiness, particularly towering cumulonimbus clouds, which may reach heights as great as 20 km (60,000 ft) and be accompanied by violent thunderstorms. A belt of thunderstorm activity occurs along the ITCZ, and the greatest activity occurs over Africa and South America. The region around Kampala in Uganda has an average of 262 thunderstorm days per year, and that around Lake Maracaibo, in Venezuela, about 297.

The ITCZ is also involved in the genesis of tropical cyclones (discussed more fully in Chapter 7), because of changes in the wind speed and direction. This is particularly the case when it moves away from the equator, and the Coriolis Effect (see Chapter 2) becomes stronger and more effective in inducing rotation of the winds. Waves move westward along the ITCZ, causing an increase in thunderstorm activity. These tropical disturbances become tropical depressions, which in turn become tropical storms and eventually tropical cyclones.

The monsoons

Many regions within the tropics tend to experience two seasonal regimes: a dry season and a wet (rainy) season. This is particularly marked in areas that are affected by monsoon conditions. In these regions the shift in location of the ITCZ is accompanied by major

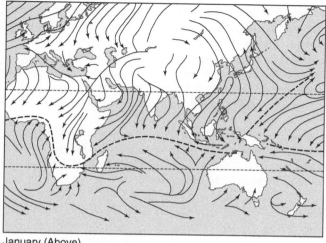

January (Above)

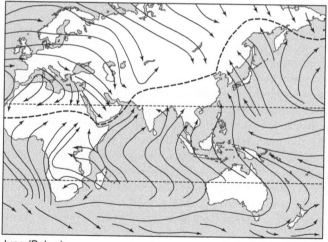

June (Below)

39. The major changes occurring in the Asian monsoons, showing the shifts in wind directions between January (above) and June (below).

changes (reversals) in the direction of the prevailing winds. The ITCZ is often then referred to as the 'Monsoon Trough'. The most extreme cases are exhibited by the West African and Asian (or Asian-Australian) monsoon systems (see Figures 12 and 39). Monsoons have been described as large-scale sea and land breezes, because they are ultimately the result of extreme differences between the temperatures prevailing over the sea and land.

In West Africa, as the ITCZ shifts northward with the onset of summer, the relatively dry north-easterly trade winds are replaced by southerly warm, humid winds from the Gulf of Guinea, heralding the start of the rainy season.

There are various, slightly different monsoon regimes that affect much of Asia. The Indian Monsoon (the South-West Monsoon) arises because a low-pressure area is created by the extreme heating over the Thar Desert in western India, as well as over the centre and north of the country. The rising air is replaced by humid air from the Indian Ocean. Some of this air actually originated as the south-east trades, south of the equator. When it crossed the equator, the Coriolis Effect caused it to deviate to the right, giving rise to south-westerly winds which then combine with the flow of air from the Arabian Gulf. When the South-West Monsoon reaches the southern portion of the Indian Peninsula, it divides into two arms. The western (Arabian Sea) arm brings heavy rainfall to the west coast (the Western Ghats). The eastern (Bay of Bengal) arm tracks north-east over the Bay of Bengal, deriving still more moisture from the sea. When this branch of the monsoon arrives at the eastern Himalaya it produces one of the wettest places on Earth in the Khasi Hills in Meghalaya, where the hamlets known collectively as Mawsynram have an average annual rainfall of 11,871 mm. (Neighbouring Cherrapunji holds the record for rainfall during one year of 26,462 mm, although this was recorded between 1860 and 1861.) After striking the eastern Himalaya, the Bay of Bengal branch of the monsoon turns westwards, following the mountain range, bringing monsoon rains to the Indus and Ganges plains. The Himalaya act as a barrier to both branches of the Indian Monsoon, causing the air to rise, cool, and produce heavy precipitation. Some areas of India may receive as much as 10,000 mm of rain annually, most of it from the South-West Monsoon.

These summer monsoon winds are replaced by the North-East Monsoon in winter. The Indian land mass cools rapidly producing a high-pressure area over northern India. As the ITCZ retreats towards the south, cold, dry air surges down from the Himalaya and the northern plains. Later in the year this flow is enhanced by a certain amount of cold air originating in the Siberian High that crosses the Tibetan Plateau. On the eastern side of India, this North-East Monsoon picks up considerable moisture from the warm waters of the Bay of Bengal, and so these winds bring considerable amounts of precipitation to the eastern side of the Indian Peninsula, which receive little rainfall from the South-Western Monsoon.

Agriculture in India is utterly dependent on the rainfall brought by the South-West Monsoon, quite apart from the relief its onset brings from the enervating high temperatures in spring and early summer. There was a major famine in India in 1899–1900, caused by the failure of the monsoon rains, and this inspired Sir Gilbert Walker to begin his investigations into the monsoon, which eventually led to his discovery of the long-range interconnections between conditions in widely spaced regions of the Earth, particularly the zonal circulation in the tropics, and to what is now known as the Walker Circulation. (Now known as 'teleconnections', such long-distance connections are discussed in Chapter 9.)

The East Asian summer monsoon produces a major shift in the location of the ITCZ to the north over the north-western Pacific and causes a significant flow of warm, moist air on generally south-easterly winds over most of Japan, the Korean peninsula, and mainland China. In the winter, the region is dominated by

the cold, dry air flowing out of the Siberian High. The ITCZ
has then retreated to the south and lies in a generally south-west
to north-east line over the western North Pacific, well to the
south-east of Japan.

Although the alternating conditions over northern Australia are
sometimes referred to as monsoon conditions, with maximum
rainfall in the southern summer, they are actually primarily the result
of the interaction of the winds with the topography (particularly that
of Borneo, which causes north-easterly winds to be diverted into
north-westerly or westerly winds directed towards Australia) and the
winds' interaction with sea and land temperatures.

Somewhat similarly, there is no true reversal of wind directions
in the conditions known as the North American Monsoon. Here
there is a flow of warm, humid south-easterly air in summer from
Mexico that affects the south-western states and may reach as far
as the ranges in southern California.

Dust storms

Major dust storms often originate in the tropical zone—or more
properly from the regions beneath the subtropical anticyclones—and
in particular from the Sahara. Although there are significant,
localized dust storms that affect countries in the region, notable
amounts of dust are often raised high into the atmosphere.
Southerly winds frequently lift Saharan dust to moderate altitudes
and carry it north over the Mediterranean and affect countries in
Europe. It is usually swept from the atmosphere by rainfall and
covers objects on the ground with 'red rain'.

A far more significant movement of plumes of dust (which may be
described as solid aerosols) from the Sahara concerns those that
are carried west by the north-easterly trade winds across the
Atlantic, and which may be deposited, not only on the ocean itself,
but as far west as the Caribbean, southern North America

(particularly Florida), and South America. This mineral dust is an extremely important source of nutrients largely responsible for the great fertility of the Amazonian forest. Further north, the Gobi Desert (which lies outside the tropics, of course) is the source of major dust storms that are carried over China by the strong westerlies. The Gobi lies in the rain shadow (see Chapter 4) of the immense range of the Himalaya, which prevent moisture-bearing air from reaching the region.

Chapter 7
Severe and unusual weather events

Although depressions may be accompanied by extreme winds and major rain- or snowfall, other events may be responsible for very severe weather. Convective clouds, for example, may range from relatively innocuous showers to major supercell systems that are accompanied by dramatic conditions, including highly destructive tornadoes.

Showers

Although, to the general public, the word 'shower' tends to suggest a short period of heavy rain, meteorologists apply the term to rain from convective cloud; either deep cumulus (cumulus congestus) or cumulonimbus, and not from extensive frontal cloud such as nimbostratus. The exact form of convective cloud is strongly dependent on the time of year. In winter, the freezing level is relatively low, and precipitation begins by glaciation (the 'cold' process) in the upper levels of the cumulonimbus cloud. But any clouds are shallow, so their water content is limited and precipitation will initially be in the form of small ice crystals, pellets, or snowflakes, which may melt into rain as they descend.

In summer, with higher temperatures, convection is much more vigorous and the clouds are deeper with a much greater water content. The freezing level is also much higher. Precipitation may be initiated by coalescence (the 'warm' process) within deep cumulus congestus clouds, but if the clouds grow up to the freezing level, the water droplets tend to become supercooled. At about –10 °C, freezing nuclei start to act, creating small ice crystals. At –40 °C the supercooled droplets freeze spontaneously. When liquid droplets collide with ice crystals they freeze into pellets of hail. Small pellets may be swept upwards by strong upcurrents, passing through layers with either liquid water or freezing conditions. If air is trapped between freezing droplets, an opaque layer results, whereas in a layer above freezing the water spreads into a layer that then forms clear ice when it freezes. Hailstones may thus consist of alternating layers of clear and opaque ice. They grow until too large to be supported by the updraughts when they fall to the surface.

Although the duration of convective cells depends on the exact conditions prevailing in the atmosphere at the time, on average each of the first two stages in the formation of a cell within a shower cloud lasts about 20–30 minutes. There may be a few large raindrops in the earliest stage as the cell is starting to develop, but most precipitation occurs in the mature stage when it reaches high levels, starting as large raindrops but then possibly changing to heavy rain accompanied by hail. However, the powerful downdraughts that develop quench the updraughts and cut off the supply of warm, humid air. This final, decaying stage may last from about 20 minutes to two hours as the rain lessens and individual drops become smaller. This means that the average, overall lifetime of a cumulonimbus cell is about 90 minutes.

When the gradient wind is strong, showers tend to be short-lived, but may be frequent during the day. When the gradient wind is weak, showers and shower clusters may last longer and become more persistent. Warm, humid air ahead of a cell is drawn towards it by the strong updraughts. The downdraughts also present in the cloud fan out when they hit the ground to create a gust front ahead of the cell. That outflowing cold air often

40. A cluster of cumulonimbus cells at various stages of development. The oldest, most distant cell has spread out into a characteristic anvil shape.

undercuts the warm air and helps to lift it into the growing cell. This flow of warm, humid air frequently leads to the formation of a daughter cell that then grows and takes over the activity as the initial cell decays. In this way a whole cluster of cells may arise, prolonging activity and bringing rain or hail to a larger area of the surface (Figure 40).

In both winter and summer, the convection may well be strong enough to reach the tropopause—which will generally be much lower in winter than in summer—and the accompanying inversion, which will arrest upward growth and cause the cloud to spread out into a layer of thick cirrus, giving rise to the characteristic 'anvil' shape (Figures 21 and 40). When convection is particularly vigorous, the top of rising cells may even penetrate a short distance into the stratosphere, giving rise to 'overshooting tops'. These are readily visible in many satellite images or photographs taken by astronauts, and may sometimes be seen from the ground above distant cumulonimbus anvils.

Thunderstorms

Large, active cumulonimbus clouds often develop into thunderstorms. Although the electrification process is still poorly understood, it seems that both water droplets and ice crystals must be present, and that cloud-top temperatures must be below −20 °C. High in the cloud, ice particles fragment on freezing, and the lighter particles become positively charged. They are carried to the very top of the cloud by the updraughts, while the heavier, negatively charged particles accumulate at the base of the cloud. There, they induce an opposite, positive charge on the ground beneath. This positively charged area drifts across the landscape beneath the cloud, until eventually the charge difference becomes so great, or the distance between the two charged areas becomes so small—above a tall object such as a building or tree—that a discharge occurs between the two. Such cloud-to-ground strokes are those most commonly visible, and are often referred to as 'fork lightning'.

There are several stages to a discharge, with the initial channel being opened by what is known as a 'stepped leader', which makes multiple short zig-zag surges towards the ground. After contact is made the main current actually passes upward from the ground to the cloud. Discharges may also occur within the cloud itself (intracloud lightning), when the discharge column is often hidden by the cloud, giving the form commonly known as 'sheet lightning'. In addition, discharges may occur between two separate clouds (intercloud lightning). The triggers for these two last forms of lightning are poorly understood, as is the mechanism behind 'bolts from the blue', where lightning discharges may extend horizontally many kilometres away from the cloud into clear air before turning down to earth. Such distant strokes do definitely exist and are an additional reason for extreme caution when lightning activity is present. Occasionally the positive charge at the top of the cloud becomes so great that a channel arises directly between it and the ground. The current in such 'positive' discharges is much greater than that in normal cloud-to-ground strokes.

The channel of air formed by a discharge is extremely hot and expands and then collapses at supersonic speeds, giving rise to the familiar thunder. The time that elapses between seeing the flash and hearing the thunder may be used to estimate the distance of the lightning discharge. A delay of three seconds equates to about one kilometre (five seconds to one mile). Sometimes flashes may be seen, but no thunder is heard. (Such a discharge is often described as 'heat lightning', mistakenly believed to occur in summer, but merely arises because the cell is too distant for thunder to be heard.) The active cell is then probably about 25–30 km away.

Although electrical activity within a single cell normally lasts 20–30 minutes, the creation of new cells may lead to a cluster of active cells, creating what is known as a multicell storm (Figure 41) that has a far greater lifetime that may amount to several hours.

41. Lightning over the sea, photographed from Sawtell, New South Wales. The distant lightning stroke shows that this was a multicell storm with at least two active centres.

Larger convective systems

Convective clouds may become organized into a large cluster of major activity. These systems, known as mesoscale convective systems (MCS), may consist of deep cumulus (cumulus congestus) or cumulonimbus, often with associated layer clouds. There is a vigorous circulation and heavy precipitation, both of which tend to become stronger as the system moves across country. Often the tops of the clouds merge to create a giant cirrus shield above the system, which may persist for four hours or more. When such an MCS exhibits a strongly linear or curved form, it may be described as a squall line. New cells are generated by the strong outflow of air, ahead of the advancing line, thus perpetuating the overall system. Squall lines may sometimes be so strong that there is a rise in pressure where there is heavy precipitation, and a drop behind the line itself.

Occasionally, one mesoscale convective system will merge with another to form a giant mesoscale convective complex (MCC). Such giant systems are actually defined technically on the basis of their properties determined from infrared satellite observations. An area that has a temperature below −50 °C must be greater than 50,000 km², and a larger area, extending over at least 100,000 km², must have a temperature below −32 °C, and the whole system persist for at least six hours. Such systems tend to occur late in the day and usually persist into the following day. They are most common over North America, Africa, and Asia and, depending on their location, may act as a precursor to the formation of a tropical cyclone.

Supercells

An even more active and dramatic form of system is known as a supercell. They arise when there is a very deep pool of unstable air accompanied by a strong increase in wind speed with height,

42. The edge of a supercell storm over Boambee, New South Wales. The curved bands of cloud are characteristic of those found in a rotating mesocyclone.

together with directional wind shear. Rather than a collection of individual cells, the circulation becomes organized into a single, large, rotating column of rising air, known as a mesocyclone (Figure 42). This column may be anywhere between 2 and 20 km in diameter, and rise as high as 8–15 km. It is accompanied by an intricate system of up- and downdraughts, and a flow of cool air entering the system at middle altitudes. Unlike normal cumulonimbus cells, where the downdraughts tend to quench the updraughts, eventually causing the system to decay, in supercells the main updraught is tilted and the rotation of the whole system separates the upward and downward flows. The whole system becomes very persistent, lasting six hours or more. At middle latitudes, supercells are most frequent in summer when there are strong temperature and humidity contrasts and are particularly common over the central and eastern states of the USA, where there is a strong contrast in the properties of air from the Gulf of Mexico and drier air to the north and west.

Supercells generally develop a giant 'vault' where the updraught is strongest. This is the ideal site for the formation of large hailstones, which may circulate up and down several times, supported by the strong updraught, and accrete numerous layers of ice before becoming so heavy that they eventually fall out of the cloud.

In addition to damaging hail, supercells frequently produce torrential rain and multiple lightning strokes. Even the powerful downdraughts have been known to cause direct damage on the ground, but, most significantly, supercells create destructive tornadoes.

Tornadoes

Tornadoes are produced by supercell systems, which, as just described, are particularly prevalent over the central states of the USA (the High Plains). Tornadoes tend to form from supercells along the dryline between the warm, moist air from the Gulf of Mexico and the drier air from the desert states to the south-west. Although details are still obscure, the method of tornado formation differs from other rotating vortices, which may be grouped under the general term of 'whirls', and which will be described shortly. Although the media tend to describe all rotating vortices as 'tornadoes' or 'twisters', there are two distinct mechanisms at work. Tornadoes appear to originate within a supercell system as a horizontal, rotating cylinder of air. The strong updraughts lift the centre of the cylinder into an arch, and the 'descending' limb with clockwise rotation decays, leaving the 'ascending' branch as the incipient tornado. (The most common direction of rotation in a tornado is thus anticlockwise at the ground, but clockwise rotation is occasionally observed. Such vortices appear in a slightly different location, relative to the parent supercell.) The separation between the up- and downdraughts in a supercell helps to strengthen the vortex, which descends toward the ground as a funnel cloud (or 'tuba'), becoming classified as a tornado when it touches down and raises a debris cloud.

The exact conditions within a tornado vortex are unknown, because, whenever encountered, all meteorological instruments have been destroyed, but the pressure drop within the column is reliably estimated at 200–250 hPa, which causes immediate condensation of water vapour in the air, giving rise to the visible funnel. Many tornadoes have diameters of approximately 100 m, but giant ones have been known to reach a diameter of 1000–2000 m. Wind speeds in tornadoes are extremely high, but again difficult to determine unless the tornado can be measured from a distance by the Doppler radar technique. The highest speed recorded is that of 514 kmh^{-1} found for the highly destructive tornado that hit the outskirts of Oklahoma City on 3 May 1999. Typical ground path-lengths are 10–100 km, although the record is held by the 'Tri-State tornado' of 26 May 1917, which caused destruction along a path of 472 km. The parent supercell of the destructive tornadoes that hit Tuscaloosa and Birmingham, Alabama, on 27 April 2011 was tracked for more than seven hours, covering more than 610 km.

The intensity of tornadoes is now judged on the Enhanced Fujita scale, originally introduced by the tornado expert Tetsuya Fujita in 1971, and was based on an assessment of the damage caused by a tornado. The initial Fujita scale, although used for many years, had shortcomings and the Enhanced scale was formally adopted in 2007. This is based on estimated (not measured) wind speeds, themselves based on damage criteria and 3-second gust speeds. Both the original Fujita scale and the Enhanced Fujita scale are given in Appendix C.

A somewhat different scale of tornado intensity has been introduced by the British Tornado and Storm Research Organization (TORRO). The TORRO scale is based on wind speed rather than on damage criteria. The TORRO scale is also given in Appendix C.

Lesser whirls

There is a whole family of vortices that are less destructive than tornadoes. They arise from completely different mechanisms. The simplest are the whirls often known as 'devils' that are produced when the wind is funnelled into a whirling column of air by the surroundings. Everyone is familiar with the way whirling columns of leaves or litter may arise when the wind is diverted by nearby buildings and similar conditions may be caused by a narrow cleft in a hillside (for example). This form is generally known by the material raised from the surface and gives rise to water, snow, and even hay devils.

Convection caused by strong heating of the surface boundary layer, especially in arid regions, may raise small particles in a column to give a dust devil, and differing roughness of areas of the surface may help to induce rotation. Generally the column peters out in mid-air, but on rare occasions, dust devils may rise high enough for condensation to create a small cumulus cloud at the top. Although dust devils generally travel across the countryside—and many, identical to those on Earth, have been observed moving across Mars—they are rarely strong enough to cause any significant damage.

Convection, but this time arising in very active clouds, is the principal mechanism that creates waterspouts and the corresponding landspouts—a term introduced by the tornado expert Howard Bluestein in 1985, because of their similarity to waterspouts—with the latter often being incorrectly described as tornadoes. In these cases, extremely strong up- and downdraughts within the cloud initiate a rotating column of air, which extends downward from the base of the cloud as a funnel cloud or tuba. Such funnel clouds are surprisingly common, and may even be produced when the sky appears to be covered with a layer of stratocumulus cloud, which is actually hiding the active cumuliform cloud from being seen from the ground. Very often such a funnel cloud does not make contact with the surface, but if it does, it becomes a landspout if over the land, and a waterspout over the sea. Waterspouts are more liable to form if there is a strong temperature contrast between warm water and much cooler overlying air. Both water- and landspouts consist of a strong downdraught, within which condensation produces the visible funnel, surrounded by a rotating, tubular updraught, usually invisible until the funnel has touched down and a small amount of material is raised into the air. In the case of waterspouts, the point at which contact is made is known as the 'dark spot'. At wind speeds above about 80 kmh^{-1} a cylinder of spray, called the 'bush', is produced. Multiple waterspouts are often recorded and similar landspout clusters probably occur, but are more difficult to detect over land where buildings and features in the landscape interfere with visibility. Landspouts may also arise from intense heating in the surface boundary layer, causing a rotating column of rising air that then links into an updraught in a convective cloud above.

Both water- and landspouts generally have diameters of 15–30 m and a short lifetime of some fifteen minutes. Waterspouts often decay if they cross onto land, although a few may continue their track as landspouts.

Very similar phenomena are often generated by the violent winds and strong convection associated with violent gust fronts and especially with the activity in tropical cyclones. In general these whirls do not generate visible funnels, but they may cause considerable damage on land. These 'gustnadoes' are often reported as 'tornadoes' or 'twisters' by the general media.

Tropical cyclones

Tropical cyclones are known as hurricanes over the Atlantic and eastern Pacific Ocean (off California and Central America), as typhoons over the western Pacific, and cyclones over the Indian Ocean. Although often regarded simply as disasters—and the accompanying high winds, extreme rainfall leading to flooding, land- and mud-slides, and storm surges may bring major death and destruction—the rainfall from these systems is vital to much tropical agriculture.

A tropical cyclone is a closed, non-frontal, low-pressure system of high winds, with maximum speeds in excess of 33 ms^{-1} (about 120 kmh^{-1}). The central pressure is often lower than 950 hPa—the current record is 870 hPa in supertyphoon Tip in 1979. Unfortunately there is no accepted international definition of when a system should be classified as a tropical cyclone, nor agreement on a scale of intensities. The World Meteorological Organization recommends the use of the ten-minute average of sustained wind speeds, measured at 10 m (the standard height for official anemometers). Overall, there are no less than five different scales used by various meteorological agencies and applied to different oceanic basins. However, the Saffir–Simpson hurricane scale (given in Appendix C), used for systems in the Atlantic, Central Pacific, and Eastern Pacific, uses the one-minute average at the same height. It has five intensity categories. Hurricane Andrew (Figure 43) was an extremely strong hurricane (Category 5 on the Saffir–Simpson scale) that struck the Bahamas and Louisiana, but caused extensive devastation in southern Florida.

The structure of all tropical cyclones shows specific features. There are bands of extremely deep convective clouds that spiral (anticlockwise in the northern hemisphere) in towards the centre where, in the strongest systems, there is a cloud-free eye, where air is actually descending towards the surface. The eye is surrounded by a band (the eyewall) of towering clouds, extreme winds, exceptionally high precipitation, and thunderstorm activity. The air, sinking in the eye, warms by descent and produces a cloud-free region that is typically 10–50 km in diameter, but may reach 70 km in the strongest systems.

43. Hurricane Andrew on 23 August 1992, when passing over the Bahamas and approaching Florida as a Category 5 hurricane. The eye is distinctly visible. The maximum gust reported for Florida was 282 kmh^{-1}.

The towering cumulonimbus clouds may attain heights of 40,000 ft (12 km) or more and the air flowing out of the system at altitude (clockwise in the northern hemisphere) creates a vast cirrus shield that may be many thousands of kilometres in diameter. (The cirrus shield above Hurricane Gilbert in 1988 had a diameter of 3500 km.)

Tropical cyclones form under very specific circumstances. They arise some 5–10° away from the equator, where the Coriolis acceleration (zero at the equator itself) promotes their overall rotation. The sea-surface temperature must be at least 27 °C, and any vertical wind shear throughout the depth of the troposphere must be very low. This would otherwise prevent the closed circulation from forming. All tropical cyclones are low-pressure systems with a warm core, unlike depressions which have a cold core. The extreme heating arises from the release of latent heat in the towering clouds that surround the centre.

There is a sequence of stages leading to the formation of a tropical cyclone. To take Atlantic hurricanes as an example, they often begin as what is known as a tropical wave (also called an easterly wave), which is a shallow, upper-air trough that moves west in the trade-wind zone above West Africa, producing an increase in cloud cover and precipitation. This then develops into a tropical disturbance, an area of organized convection and associated with weak low pressure. Subsequently, this turns into a tropical depression, a low-pressure area with closed isobars and circulation. Wind speeds are relatively low, being less than 18 ms^{-1}. (This is about Force 7 on the Beaufort scale.) These systems tend to originate at the Intertropical Convergence Zone, and many do not develop any further. When the wind speeds rise and curved cloud bands become visible in satellite images, the system has become a tropical storm, and usually receives a distinctive name at this stage. Wind speeds may rise to 26–32 ms^{-1} (Force 10–11). Finally, with still higher wind speeds the system becomes a fully fledged hurricane.

The tracks of tropical cyclones may be erratic, but in general they move westward (Figure 44) and slowly track towards the poles at about 10 knots (19 kmh^{-1}). If they reach latitudes of 20–30° N or S they often exhibit a dramatic change in direction (known as recurvature), swinging round to the north-east or south-east (in the northern and southern hemispheres, respectively). Hurricane Sandy, which caused extensive destruction to New Jersey and New York in 2012, was an exception to this behaviour. After tracking northward offshore along the East Coast of the United States, it encountered a high-pressure system, located over New England. Sandy abruptly curved to the north-west and made landfall, albeit as a Category 1 hurricane (the weakest) on

111

44. Typical tracks of tropical cyclones. Note the sudden, major change of track (recurvature) when they move away from the tropics.

Arctic Circle

Tropic of Cancer

Equator

Tropic of Capricorn

29 October 2012. The accompanying storm surge caused extensive flooding of streets, tunnels, and the subway system in New York city, quite apart from extensive wind damage.

At the late, recurvature stage, most systems start to decay as they move away from warm water, and especially if they move over land and lose their source of heat. They may then continue to higher latitudes, either as extratropical cyclones in themselves, or merge with existing low-pressure systems—usually causing the latter to deepen dramatically.

Chapter 8
Localized weather

Although certain of the events described earlier, such as violent
tornadoes, affect relatively small areas on the ground, there are a
number of effects that are localized in their influence. Consider fog.
This may be associated with widespread anticyclonic conditions
which lead to a significant drop in temperature at night, and
relatively quiet, or windless, conditions. There are two common
form of fog: radiation fog, and advection fog.

Radiation fog

Radiation fog occurs when the surface radiates heat away to
space, cooling low-lying air below the dewpoint. It is frequently
confined to specific areas, such as the immediate vicinity of
streams or narrow river valleys. Several conditions are required
for cooling to take place:

- The sky must be clear, allowing long-wave radiation to escape
  to space
- There should either be no wind, or only a very light wind of
  less than four knots, 7.5 kmh^{-1}
- The air must be humid in late evening. (This condition is most
  often met on autumn and winter evenings.)

- There should be sufficient time for the air to cool to the dewpoint. (Once again, this condition is most likely to be met in autumn and winter.)

Radiation fog is usually confined to the lowest areas of the land, but even so, generally it is confined to a layer that is between 15 and 100 metres deep. Occasionally, especially after heavy rain that leaves the air very humid, much thicker layers may form, particularly in river valleys or near lakes or reservoirs, to produce heavy valley fog (Figure 45).

A moderate wind will tend to prevent fog forming, because the associated turbulence mixes a deep layer of air and inhibits cooling. Similarly, if the wind rises after fog has formed, it will tend to disperse for the same reason.

If the temperature falls low enough, fog that has condensed as dew onto objects on the ground will freeze to create a layer of hoar frost. If, however, the fog droplets become supercooled, any slight drift of air will cover objects on the ground with a coating of rime, where the supercooled droplets have frozen the instant they came in contact with a solid surface. Similar rime occurs at higher altitudes when supercooled cloud droplets may form long 'feathers' on the windward side of objects such as aerial masts. Such 'feathers' grow into the wind, rather than downwind.

Fog tends to disperse in the morning after there has been sufficient heating for gentle convection to set in. The layer of fog will often lift away from the surface and form a layer of low stratus cloud before finally breaking up and dispersing. Sometimes, scattered remnants of fog may be seen moving up the sides of valleys or mountains, carried by the local flow of air. In river valleys, air flowing downslope and downstream will often carry the remnants of fog out towards lower-lying land.

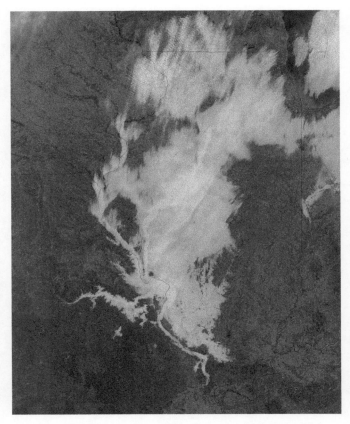

45. **Extensive fog in the valleys of the Missouri and Mississippi rivers on 26 March 2012.**

Advection fog

Advection fog forms when warm, humid air flows over a very cold surface, such as the sea or, occasionally, snow- or ice-covered land. Over the sea, in particular, extensive areas of fog may be produced, and this fog is then advected (transported horizontally) onto neighbouring coasts. With low-lying land, the fog may invade areas

116

many kilometres from the sea. Over land, such a fog may burn off during the day, persisting over the sea, but returning to the land at night when the temperature drops.

A similar type of fog often occurs when the land has been covered by snow and a thaw begins, with temperatures hovering around 0 °C. A drift of air may then carry the fog a considerable distance across the countryside.

Haze and smog

A different form of obscuration occurs when large quantities of tiny, dry particles (aerosols) are suspended in the air. The particles tend to scatter light, rendering distant objects indistinct and sometimes contributing to dramatic sunrise and sunset colours. Haze usually builds up during the day and may occur as a distinct layer with a brownish coloration.

Both dry haze and water-droplet fog may be contaminated with pollutants to give rise to photochemical smog. This may be of two distinct types depending on the nature of the pollutants although both usually contain injurious substances such as carbon monoxide. Often the major pollutants are nitrogen oxides and hydrocarbons, emitted by vehicles. This type forms the notorious Los Angeles smog, where the polluted air is trapped over the city by surrounding high land. Sulphur compounds tend to arise from widespread combustion of coal and oil. It was this combination of fog and smoke that gave rise to the Great Smog of 1952 that affected London for five days in December, and produced at least 4,000 fatalities (and affected some 100,000 more with severe respiratory problems), although recent research suggests a much higher death toll.

Some major pollution events combine both hydrocarbon and sulphurous compounds. This is particularly the case with the major pollution events that affect China, where the large numbers of coal-burning power stations emit vast quantities of pollutants.

Some events are so severe that the levels of pollutants are over forty times the maximum amount recommended by the World Health Organization.

Both types of smog are, of course, like fog, often confined to low-lying land such as river valleys and relatively enclosed basins. This is often clearly revealed by satellite images of the areas concerned.

Local winds

There are a number of relatively localized winds that may have a distinct effect on the weather in a particular area. Five are sufficiently distinct to be described individually. These may be considered as two groups with similar mechanisms. One includes the sea, land, and lake breezes. The second links valley and mountain winds.

Sea, land, and lake breezes

During the day, the land warms more rapidly than any adjacent area of water, and a shallow circulation develops, with warm air rising over the land and cool air flowing in from the sea (in the case of a sea breeze) to replace the rising air. Such a breeze may bring sea fog over the coast, so a warm fine morning is replaced by a misty cool afternoon. Sea breezes are most common on the coast in spring and early summer when the sea is relatively cold. They tend to set in before midday, and reach a maximum speed during the afternoon. There is often a distinct sea-breeze front, which may be accompanied by a line of cumuliform cloud, especially where a line of hills inland may provide an extra lift to the rising air and give rise to rain or, very occasionally, even thunderstorms. Sea breezes may penetrate many tens of kilometres inland.

When there is a peninsula, sea breezes may form from both sides of the land and propagate towards one another. Where these sea breezes meet, especially over a higher spine to the peninsula, significant cloud and precipitation may occur. This has been the cause of major downpours and flooding such as the dramatic flooding in Boscastle in Cornwall in 2004. On a much larger scale, sea breezes from both sides of the Cape York Peninsula in Queensland, Australia, sometimes interact to produce the spectacular roll cloud known as the Morning Glory, which propagates westward across the Gulf of Carpentaria. This is accompanied by a sudden jump in pressure and may appear as a sequence of separate roll clouds.

The land breeze is the counterpart of the sea breeze. It sets in when the land cools (more rapidly than the sea) at night, resulting in an outflow of cool air over the sea. As with the sea breeze, there may be a distinct land-breeze front of cumuliform cloud that gradually moves farther out to sea and this may often be seen on satellite images.

A somewhat similar mechanism to that of the sea breeze may produce a lake breeze, where cool air from a large body of water, such as a lake or reservoir, invades the neighbouring land. In this case, the effect may be complicated by the surrounding topography and the orientation of the body of water and any neighbouring hills. Warming of adjacent hillsides by sunlight during the day may give an additional impetus to the flow of air which is correspondingly strengthened.

Valley and mountain winds

Heating of hill- or mountain-sides during the day creates a flow of air upslope and this may develop into a full-scale valley wind flowing up a valley towards its head. Such a wind begins to develop at sunrise and shortly after midday reaches a maximum strength, which may be 20 kmh^{-1} over sun-warmed slopes but much less over shaded areas. The strength may, of course, be modified by an existing gradient wind, and when the overall wind is strong, turbulence over rugged countryside may completely disrupt any local valley wind.

The opposite effect—cooling at night—creates a mountain wind, which begins to set in as heating ceases at sunset. The cooled air slides downhill, but the wind (often called a mountain breeze) is generally weaker than the corresponding valley wind, and may reach speeds of about 12 kmh⁻¹, but may exceed this if the valley narrows into a gorge.

The topography may, of course, sometimes have a great effect on gradient winds themselves, irrespective of valley or mountain winds. Wind speeds may increase greatly when the airflow is confined to a narrowing valley or gorge that creates a funnelling effect. Such an effect occurs in the valley of the Danube, for example, where it cuts through the Carpathian Mountains. In this case, the strengthened wind is known as the kosava. A similar effect occurs between the mountainous islands of Corsica and Sardinia in the Mediterranean, causing extreme winds that are a major hazard to sailors. A westerly wind in the Straits of Gibraltar may double in velocity as a result of being confined between the high ground on both sides of the Strait.

Katabatic winds

Winds that are very similar to mountain winds may occur when a pool of air above high ground becomes extremely cold, especially if the surface is covered in snow or ice. The dense air rushes downslope in a katabatic (or 'fall') wind, and generally affects a much larger area than ordinary mountain winds. There are a number of named winds of this sort, the most famous being perhaps the mistral that forms an icy blast down the Rhône valley and out over the Golfe du Lion. The bora that affects the eastern Adriatic and rushes down the sides of the Dinaric Alps is a similar katabatic wind, but the most extreme form occurs around Antarctica, when violent winds cascade off the East Antarctic ice sheet. The windiest place in the world is Commonwealth Bay on the George V coast with an average annual wind speed of 67 kmh⁻¹, and the greatest wind speed recorded at the same site reached 320 kmh^{-1}. The Helm Wind—the only named British wind—which affects parts of the Eden Valley in Cumbria, and cascades over the Mallerstang Escarpment, is a violent katabatic wind.

Föhn conditions

With the katabatic winds just described, the air remains cold, despite heating during its descent. Certain fall winds, however, may become extremely warm under what are described as föhn conditions. The name originates from the situation in southern Germany when air from the south descends the abrupt northern side of the Alps. As described earlier, ascending air will initially cool at the dry adiabatic rate (DALR) and then, when condensation sets in, at the saturated adiabatic rate (SALR). But if some of the water vapour is lost as precipitation (whether rain or snow) on the windward side of the hills or mountains, when the air descends to leeward it will begin to warm at the greater DALR rate sooner than otherwise might be expected. At any given level it will be warmer to leeward than on the windward side of the range. Föhn conditions may sometimes occur very suddenly, producing a dramatic rise in temperature at the base of the hills. This may be more than enough to melt any lying snow, and also often has an extreme desiccating effect, which may even create a fire hazard. Föhn conditions caused the greatest two-minute temperature rise of 27 deg. C (from –20 °C to 7 °C) recorded at Spearfish, South Dakota, on 23 January 1943.

Somewhat similar temperature rises may sometimes occur when air from middle levels of the troposphere is forced to descend to ground level on crossing a mountain range, when the barrier has created exceptionally large waves in the stream of air.

Lake effect snow

Precipitation may also be greatly localized. Although this is usually the case with rain, hail, or snow from showers, one

46. The lake effect snow was particularly deep on the southern shore of Lake Erie.

extreme form is what is termed 'lake effect snow'. This is particularly noticeable on the shores of the Great Lakes in North America, when extremely cold arctic air crosses the still-unfrozen water. In doing so, it not only gains humidity from the surface but becomes very unstable throughout a relatively shallow depth, producing vigorous shower clouds. On encountering the shoreline, these may deposit enormous quantities of snow on the surface. The area around Lake Erie and Buffalo, New York, is particularly prone to receive extreme snowfall from this source (Figure 46). Similar effects may, of course, occur elsewhere, when the source of the humidity and instability is the sea rather than a lake. (The events are then sometimes known as 'ocean effect snow' or 'bay effect snow'.) One such effect sometimes occurs when an exceptionally cold air-stream crosses the North Sea and deposits heavy snowfall over the eastern counties of Britain.

Ice storms and glaze

When liquid raindrops fall into a layer of extremely cold air, they may become supercooled, remaining liquid until they hit the surface, when they immediately freeze into a coating of clear ice. This is known technically as glaze or, more commonly, as 'black ice'. Quite apart from its effect on transport, when glaze makes road surfaces highly treacherous, the weight of ice that accumulates on objects may cause extensive damage to trees and overhead power and communications lines. Although generally confined to relatively small areas of the ground, on occasion the precipitation accompanying a major depression may affect a vast swathe of the countryside. This was the case with the 'ice storm' that hit Canada and New England in January 1998, which caused immense amounts of damage to trees and power lines, resulting in a widespread disruption to electricity supplies that lasted for weeks.

Chapter 9
Worldwide effects
and forecasting

Observing the weather for the purposes of forecasting is a worldwide enterprise. Routine measurements are made by national agencies in every country of the globe. These observations, and their distribution, are handled by international agreement and a specialized agency of the United Nations, the World Meteorological Organization (WMO), based in Geneva. All data, wherever obtained, are made freely available to all the nations that participate, and the whole system is known as the World Weather Watch (WWW). It consists of three primary elements: the Global Observing System, which ensures observations are made and reported in a standardized form; the Global Data-Processing System with standard procedures for the reception, processing, storage, and retrieval of every meteorological observation anywhere in the world; and the Global Telecommunications System which is the physical communications network covering every part of the world. About 10,000 observations are distributed around the world every hour of the day and night by the Global Telecommunications System.

Many observations are now obtained by automatic weather stations (AWS), so almost continuous data may be available. Nevertheless, the WMO stipulates that certain key observations should be made simultaneously, everywhere, at certain fixed times. These are always taken on the hour as given by Coordinated Universal Time

(UTC), which corresponds to the time on the Greenwich meridian, and does not pay any regard to changes to civil time for Summer Time or Daylight Saving Time. (UTC itself is obtained by the intercomparison of several, highly accurate, atomic clocks maintained at various standards laboratories around the world.) The time zone centred on the Greenwich meridian is designated 'Zulu', abbreviated 'Z', and it is common for meteorological observations to be given as obtained at '00:00 Z', for example. Two key times for observations are stipulated, 00:00 Z and 12:00 Z, although many stations make observations at other times during the day, or even every hour.

Apart from observations from land-based stations, ships at sea, aircraft in flight, and meteorological satellites, observations are also obtained from radiosondes, instrumented packages that are raised through the atmosphere by balloons. Such radiosondes may not only gather information about temperatures, humidities, and pressures as they ascend, but in certain cases may also monitor ozone or radiation concentrations. Tracking of a radiosonde—normally, nowadays, by using GPS data—allows the direction and velocity of winds at different levels in the atmosphere to be determined. Again, the release of radiosondes occurs worldwide at specific times laid down by the WMO (generally twice a day), to ensure compatibility with other observations obtained by different stations.

Automatic weather stations are now located at sites that are almost inaccessible or subject to extreme conditions, such as sites in Antarctica, on high mountain peaks, or meteorological buoys moored in mid-ocean. There are also hundreds of free-floating buoys, adrift on the world's oceans, including some sophisticated ones that sink, on a regular basis at predetermined times, to certain specific depths to obtain data such as temperature and salinity. Such buoys return to the surface to transmit their data and location, and then redescend into the ocean. Many of these remote automatic systems return their data by transmitting it to relay satellites, which then re-transmit the data when in range of land-based receiving stations.

In recent decades it has become obvious that events in one part of the world may have a profound effect on the weather experienced in distant regions. The first such major distant connection was established by Sir Gilbert Walker who, in studying the Indian monsoons, realized that there was a correlation between pressures at widely separated locations in the Indian and Pacific Oceans. When pressure rose over Tahiti, it fell over Darwin in Northern Australia. The oscillation between the pressure over the Central Pacific and that over the Indian Ocean is now called the Southern Oscillation, and its state, derived from the sea-level pressure over Tahiti minus the pressure over Darwin, is known as the Southern Oscillation Index.

Walker also established a link between rainfall over India and Java and atmospheric pressure over the Pacific, and established the existence of large-scale zonal circulations in the tropics along lines of latitude. These zonal circulations are now known as Walker cells and the overall motion as the Walker Circulation. Long-distance relationships between various atmospheric conditions are now known by the general term of 'teleconnections'.

Teleconnections

The most famous (or should one say 'notorious'?) teleconnection is El Niño, the term for the 'warm' phase of a particular teleconnection affecting the tropical eastern Pacific Ocean and now known to be part of a larger oscillation, the El Niño Southern Oscillation (ENSO). Under normal conditions, there is a high-pressure system over the eastern Pacific (near South America), where there is strong up-welling of cold oceanic water. A low-pressure system is located over Indonesia, with strong convection, and high sea-surface temperatures. The easterly winds of the Walker Circulation are relatively strong. If this circulation weakens, the El

Niño phase begins. Pressure over Indonesia increases, convection weakens, and warm surface waters move east, towards South America, eventually replacing the cold up-welling. Conversely, a strong Walker Circulation creates a 'cool' (La Niña) phase, when the sea-surface temperature in the eastern Pacific is lower than normal and pressures are high in the east and low in the west. The disruption to the normal circulation pattern is not confined to the tropical Pacific, but extends its influence to many other parts of the globe. A strong El Niño event (such as the extremely strong event occurring in late 2015) is ultimately responsible for drought in eastern and southern Africa, torrential rains, flooding, and mudslides in California, and also contributes to the delay in the onset of the rainy season over Indonesia and neighbouring areas. It is also considered to be partially responsible for the increased strength of cyclones in the Indian Ocean—two unprecedented cyclones hit Yemen in late 2015, and Hurricane Patricia, the strongest ever recorded in the eastern Pacific, devastated Mexico at the same period.

The North Atlantic Oscillation

Such effects are not confined to the tropics, and it appears that such events also affect storminess in the North Atlantic and elsewhere. The significance of these teleconnections is becoming more appreciated. It has become obvious that there are other oscillations that create cycles of weather activity in various regions. There is, for example, the North Atlantic Oscillation (also discovered by Sir Gilbert Walker), which may be most simply understood as arising from fluctuations in the relative strength of the permanent subtropical Azores High and the semi-permanent Icelandic Low. When pressure in the south is high (and correspondingly low in the north), in what is termed a high-index phase, the westerly winds increase in strength and depressions track at higher latitudes. When the index is low, the westerlies decrease, depressions track further south, and tend to affect the Mediterranean, with increased rainfall in southern Europe and north Africa. Northern European countries, by contrast, have cold, dry winters. With a low index there is also a tendency for outbreaks of extreme arctic air to affect the north-eastern United States and Canada.

The North Atlantic Oscillation is closely linked to the Arctic Oscillation (AO), which is also known as the Northern Annular Mode (NAM). This involves the pressure over the Arctic and the strength of the northern polar vortex. In the 'positive phase', a pool of extremely cold air over the Arctic, accompanied by a strong vortex and high pressure at mid-latitudes (around 45° N), shifts the track of depressions further north, giving wetter weather, not just to countries bordering the eastern Atlantic, but also to Alaska, and drier conditions across the west of North America and in the Mediterranean region. The north-easterly trade winds are stronger. Greenland, Labrador, and Newfoundland are particularly cold.

In the corresponding 'negative phase', a weak vortex accompanies depression tracks farther south (affecting the Mediterranean). There are weaker north-easterly trade winds, and cold arctic air tends to penetrate farther into Europe, and, on the other side of the Atlantic, into the midwest and eastern seaboard of North America.

There are longer-term oscillations that appear to operate on a decadal scale. The best known of these are the Atlantic Multi-decadal Oscillation (AMO), which should not be confused with the North Atlantic Oscillation just described. In contrast to the latter, which is based on atmospheric changes, the AMO is an apparent long-term fluctuation in the oceanic currents and resulting sea-surface temperatures. There is disagreement as to the significance of the changes that have been established.

The Pacific Decadal Oscillation

A somewhat similar oscillation affects the northern Pacific, known as the Pacific Decadal Oscillation, evidence for which appears to be slightly more robust than that for the AMO. (Both of these decadal oscillations seem to be the result of the combined action of several processes with different origins.) The phase of the PDO is defined from sea-surface temperature at middle latitudes of the Pacific. In its 'positive' phase the oceanic temperature in the western region becomes cooler, and warmer in the eastern. Particularly in the northern winter this leads to humid winds affecting north-western Canada and the United States, with increased rainfall. Opposite conditions apply during a 'negative' phase.

It has been established that there is a specific link between a strong El Niño event and the positive phase of the PDO. This link is known as an 'atmospheric bridge'. Enhanced convection over the pool of warm water in the eastern tropical Pacific creates planetary waves that are enhanced over the area of the semi-permanent Aleutian Low, which deepens, increasing winds speeds around it and the consequent effects on the track of depressions over the northern Pacific.

The solar influence

A correlation between solar activity and auroral events has always been particularly obvious, but for many years there have been multiple attempts to link weather at the Earth's surface to changes in solar activity. These attempts have usually taken the form of trying to establish a correlation between various aspects of the weather (cold winters, extreme storms, etc.) and sunspot numbers, which change on a roughly eleven-year cycle. (In reality this period is half of the Sun's approximately twenty-two-year magnetic cycle.) But visible sunspots are only one aspect of solar activity, and none of the attempts to find some relationship between weather and sunspot numbers has ever been successful. It is known, however, that there is a link between solar activity and the upper atmosphere. When the Sun is particularly active, there is heating of the upper atmosphere and the atmosphere actually expands. This has a direct effect on the orbits of satellites, which experience additional drag, causing the orbits to decay more rapidly. Bombardment by solar radiation may also degrade satellite electronic systems (in particular).

In recent years it has been established that there is an actual correlation between solar activity and the strength of the Arctic polar vortex (and presumably also that of the Antarctic vortex). Although the exact mechanisms at work are yet to be established in detail, enhanced strength of the Arctic vortex would, in turn, affect the North Atlantic Oscillation. It seems that greater solar activity accompanies an enhanced northern vortex and creates a stronger, zonal, polar jet stream. This in turn leads to more and stronger depressions traversing Western Europe, with the accompanying strong winds and increased rainfall. With low solar activity and a weak vortex, as discussed in Chapter 3, the jet stream exhibits greater north–south (meridional) waves, creating blocking situations and bringing more polar air south to middle latitudes, leading to much colder winters.

One direct effect of solar activity involves its output of ultraviolet radiation. This has opposing effects on the existence of ozone in the ozone layer. Not only does ultraviolet radiation create ozone by breaking apart oxygen molecules (O_2), which then combine into ozone (O_3), but the return of solar radiation (particularly ultraviolet radiation) in the polar spring creates chemical radicals that act (on the surface of the ice particles in polar stratospheric clouds) to break down any ozone that is present, leading to the ozone holes. There is an unconfirmed suggestion that fluctuations in solar activity may have an effect on the amount of depletion in ozone at any particular time.

Space weather

In 2014 the UK Met Office established the Space Weather Centre. This is concerned, not with weather in the accepted meteorological sense—although there may be links between solar activity and 'conventional' weather as just discussed—but with the impact that solar activity has on various activities and infrastructure. Specific events to be covered are solar flares, geomagnetic storms, and coronal mass ejections (CMEs). The last involve the ejection of large amounts of gas and magnetic fields from the solar corona, which have major effects if they impact on the Earth's magnetic field.

Apart from the effects on the orbital decay of satellites mentioned in 'The solar influence', there may be disruption of the accuracy and reception of signals from Global Navigational Satellite Systems (GNSS) and Global Positioning Systems (GPS). In addition, major geomagnetic storms may create (and have created) the breakdown of electricity transmission grids. Interruptions to radio communication channels through changes in the ionosphere have long been known to occur. An additional hazard to airline passengers and crew is posed by enhanced radiation at high altitudes (and high latitudes). The prediction of the likelihood of these events, and possible strategies for minimizing their effects, has become increasingly important with our growing reliance upon complex technology.

Weather forecasting

The more conventional form of weather forecasting has become increasingly sophisticated in recent years. The basic process used in the preparation of the majority of forecasts, worldwide, is numerical weather prediction (NWP) using numerical models of the atmosphere. Synoptic data—that is data obtained at the same time over a large area (for preference over the whole globe)—is used as input to the models together with other data, such as the information from satellite sensors, obtained at various times. In favourable cases, the data are obtained at multiple levels in the atmosphere. (One model used by the British Met Office currently calculates parameters at seventy levels in the atmosphere.)

Observational data are obtained from fixed, land stations, from ocean-going vessels, aircraft, and, increasingly, from satellites, both the geostationary satellites that continuously monitor a specific view of the planet and polar-orbiting satellites that, at a much lower altitude, provide regular coverage of swathes of the surface. Satellites employ what is known as 'topside sounding' and, with their increasingly sophisticated instrumentation, are able to provide data on numerous parameters—even, in the latest sensors, surface pressure—over wide areas of the Earth.

Supercomputers are then used to solve the enormous set of interdependent equations that describe the changes that occur in specific parameters (such as pressure, temperature, humidity, wind speed, and wind direction) in the model atmosphere over a period of time. The interval concerned may range from a few hours to a few days, depending on the model being employed. From the initial situation (used to prepare an analysis chart), the situation that is likely to occur at future times may be derived and used as the basis for a forecast.

It is well known (from chaos theory) that there are inevitable inaccuracies both in obtaining the data and in predicting future conditions, and that minor (seemingly insignificant) differences in the initial data may lead to major differences in the final result. Unfortunately chaos theory has become interpreted by the general public in terms of what they understand is the Butterfly Effect (see Box 6). However, chaos theory itself is turned to good effect in what is termed 'ensemble forecasting'. In this procedure, forecast calculations are carried out numerous times with small differences introduced into the initial input data. The resulting predictions are then examined. If the different calculations produce similar results, then the forecast may be regarded as fundamentally sound. If, however, the results differ widely, then the predictions are likely to be unreliable. Such a comparison therefore offers a means of determining the likely accuracy of any specific forecast.

Box 6 The Butterfly Effect

This term arises from the title of a paper given by the meteorologist Edward Lorenz in 1972 that first brought chaos theory to general attention. The title, chosen, not by Lorenz, but by the chairman of the conference he was attending, was 'Predictability: Does the Flap of a Butterfly's Wings in Brazil Set off a Tornado in Texas?' In fact, the title could well have been 'Predictability: Does the Flap of a Butterfly's Wings in Brazil Prevent a Tornado in Texas?' Lorenz's central concept was not that minor changes lead to major effects, but that the question was essentially unanswerable. Errors in weather forecasting (in particular) are inevitable and arise because of inadequate data coverage, unavoidable inaccuracies in the initial measurements, incomplete knowledge of the underlying physics, and also because the equations used for making human or computer predictions are always approximations. The term—formally known as 'sensitive dependence on initial conditioning'—is more correctly used to imply that there are limits to predictability, not that such an effect exists.

Chaos theory is a field of mathematics that studies the behaviour of mechanical or physical systems—weather being a significant example of the latter—which are apparently simple and governed by well-known physical laws. In any such system, when computations using the appropriate equations are carried out, minute variations in the initial conditions may result in wildly different end results. Regrettably, the application of chaos theory to the weather has become known (incorrectly) to the general public as 'The Butterfly Effect', with the implication that minor events have major results.

Recent years have seen an increase in the use of 'nowcasting'. This is the use of current data—i.e. data not necessarily taken at the standard synoptic times—to prepare a short-term forecast (generally for a restricted area and for a limited time, typically six hours). The data employed may include current satellite images, radar images of rainfall patterns, and the location of lightning strikes derived from their radio emissions—known as 'sferics'. Such forecasts are able to describe localized weather systems that are not particularly well determined from synoptic data. Such broadcast forecasts are frequently accompanied by a description of current weather.

The increased sophistication of modern forecasting methods has undoubtedly improved the accuracy of forecasts. It is now common for forecasts to be available for at least three days ahead, and studies have shown that modern three-day forecasts are as accurate as those given, thirty years ago, for just a single day ahead. Forecasts for even longer terms are now widely available, medium-range forecasts covering three to ten days ahead, and long-range forecasts (or 'outlooks') give probable conditions more than ten days in advance. Ensemble methods are often used to assess the likely accuracy of these longer forecasts. Such longer-term weather forecasts begin to take account of the possible effects of known teleconnections such as the ENSO system and provide some assessment of the probable impact of major changes in weather patterns that may, for example, affect agriculture in forthcoming seasons. Predictions of long-term changes in the overall climate of various regions take us into the realm of climatology and the various models that are used to predict the fluctuations that may occur because of climate change.

Appendix A

The Beaufort scale of wind speeds

The Beaufort scale (for use at sea)

The original scale applied to conditions at sea, and was described in terms of the effect upon a frigate of the period. It was subsequently modified for use on land, and all the descriptions were generalized to be universally applicable.

FORCE	DESCRIPTION	SEA STATE	SPEED	
			KNOTS	ms^{-1}
0	calm	like a mirror	<1	0.0–0.2
1	light air	ripples, no foam	1–3	0.3–1.5
2	light breeze	small wavelets, smooth crests	4–6	1.6–3.3
3	gentle breeze	large wavelets, some crests break, a few white horses	7–10	3.4–5.4
4	moderate breeze	small waves, frequent white horses	11–16	5.5–7.9
5	fresh breeze	moderate, fairly long waves, many white horses, some spray	17–21	8.0–10.7

Weather

FORCE	DESCRIPTION	SEA STATE	SPEED	
			KNOTS	ms^{-1}
6	strong breeze	some large waves, extensive white foaming crests, some spray	22–7	10.8–13.8
7	near gale	sea heaping up, streaks of foam blowing in the wind	28–33	13.9–17.1
8	gale	fairly long and high waves, crests breaking into spindrift, foam in prominent streaks	34–40	17.2–20.7
9	strong gale	high waves, dense foam in wind, wave-crests topple and roll over, spray interferes with visibility	41–7	20.8–24.4
10	storm	very high waves with overhanging crests, dense blowing foam, sea appears white, heavy tumbling sea, poor visibility	48–55	24.5–28.4
11	violent storm	exceptionally high waves may hide small ships, sea covered in long, white patches of foam, waves blown into froth, poor visibility	56–63	28.5–32.6
12	hurricane	air filled with foam and spray, visibility extremely bad	≥64	≥32.7

The Beaufort scale (adapted for use on land)

FORCE	DESCRIPTION	EVENTS ON LAND	SPEED	
			kmh^{-1}	ms^{-1}
0	calm	smoke rises vertically	<1	0.0–0.21
1	light air	direction of wind shown by smoke but not by wind vane	1–5	0.3–1.5
2	light breeze	wind felt on face, leaves rustle, wind vane turns to wind	6–11	1.6–3.3
3	gentle breeze	leaves and small twigs in motion, wind spreads small flags	12–19	3.4–5.4
4	moderate breeze	wind raises dust and loose paper, small branches move	20–9	5.5–7.9
5	fresh breeze	small leafy trees start to sway, wavelets with crests on inland waters	30–9	8.0–10.7
6	strong breeze	large branches in motion, whistling in telephone wires, difficult to use umbrellas	40–50	10.8–13.8
7	near gale	whole trees in motion, difficult to walk against wind	51–61	13.9–17.1
8	gale	twigs break from trees, difficult to walk	62–74	17.2–20.7

Appendix A

FORCE	DESCRIPTION	EVENTS ON LAND	SPEED	
			kmh^{-1}	ms^{-1}
9	strong gale	slight structural damage to buildings; chimney pots, tiles, and aerials removed	75–87	20.8–24.4
10	storm	trees uprooted, considerable damage to buildings	88–10	24.5–28.4
11	violent storm	widespread damage to all types of building	102–17	28.5–32.6
12	hurricane	widespread destruction, only specially constructed buildings survive	≥118	≥32.7

Appendix B
The major cloud types (genera)

Cloud type	Étage (level)	Description
altocumulus	middle	a layer of individual cloudlets, showing shading
altostratus	middle	a layer of featureless cloud; Sun may be visible as through ground glass
cirrocumulus	high	a layer of small, individual cloudlets, showing no shading
cirrostratus	high	a thin layer of ice crystals, often with optical phenomena
cirrus	high	wisps of ice crystals, occasionally dense enough to appear grey
cumulonimbus	all levels	towering convective cloud, often the source of heavy rain, hail, or lightning
cumulus	low	individual, 'fluffy' clouds
nimbostratus	middle to low	dense rain-bearing cloud in depressions, often lowering near to ground

Cloud type	Étage (level)	Description
stratocumulus	low	a layer of large, grey, individual cloudlets, with breaks allowing rays of sunshine to penetrate or blue sky to be visible
stratus	low	a continuous, unbroken layer of grey cloud

Appendix C

Tornado and hurricane intensity scales

Fujita scale and Enhanced Fujita scale of tornado intensities

The original scale for describing the severity of a tornado or other severe wind was based on the intensity of damage that is observed. The maximum wind speed is estimated from an analysis of the damage, and is thus not directly comparable with direct measurements. Class F5 tornadoes are rare.

Fujita scale (Fujita–Pearson scale)

SCALE NUMBER	WIND SPEED		DAMAGE
	mph	kmh^{-1}	
F0	≤72	≤116	light
F1	73–112	117–80	moderate
F2	113–57	181–251	considerable
F3	158–207	252–330	severe
F4	208–60	331–417	devastating
F5	≥261	≥418	incredible

Because of shortcomings in the original scale, including failure to consider different types of building construction and the inability to classify a tornado if no damage was observed, the Enhanced Fujita scale was formally introduced in February 2007. Note that the scale is still based on wind-speed estimates (not measurements). Both scales are specified in non-metric units.

Enhanced Fujita scale

EF-SCALE NUMBER	WIND SPEED (3-sec. gust)	
	mph	kmh^{-1}
EF0	65–85	105–37
EF1	86–110	138–77
EF2	111–35	178–217
EF3	136–65	219–66
EF4	166–200	267–322
EF5	≥200	≥322

The TORRO scale of tornado intensities

A scale of tornado intensity, developed by the Tornado and Storm Research Organization (TORRO) in the United Kingdom. It is defined on the basis of wind speed, rather than on the intensity of damage.

TORRO scale

SCALE NUMBER	WIND SPEED		NAME
	ms^{-1}	kmh^{-1}	
T0	17–24	61–86	light
T1	25–32	90–115	mild
T2	33–41	119–48	moderate
T3	42–51	151–84	strong

SCALE NUMBER	WIND SPEED		NAME
	ms^{-1}	kmh^{-1}	
T4	52–61	187–220	severe
T5	62–72	223–59	intense
T6	73–83	263–99	moderately devastating
T7	84–95	302–42	strongly devastating
T8	96–107	346–85	severely devastating
T9	108–20	389–432	intensely devastating
T10	>121	>436	super

Note that the scale is defined in terms of wind speeds in ms^{-1}, but that the speeds in kmh^{-1} are rounded conversions, and thus appear discontinuous.

Saffir–Simpson hurricane scale

The scale was originally developed for describing the potential severity of Atlantic hurricanes, but was subsequently applied to all tropical cyclones in the Atlantic, Central Pacific, and Eastern Pacific basins. The assessment is in terms of both the wind speed and possible storm-surge height. Note that the scale is actually defined in non-metric units, so the metric figures are conversions.

Saffir–Simpson scale (Saffir–Simpson damage potential scale)

CATEGORY		CENTRAL PRESSURE		WIND SPEED		STORM SURGE	
		in	hPa	mph	kmh^{-1}	ft	m
1	weak	>28.94	>980	74–95	104–33	4–5	1.2–1.5
2	moderate	28.50–28.91	965–79	96–110	134–54	6–8	1.8–2.5
3	strong	27.91–28.47	945–64	111–30	155–82	9–12	2.8–3.7
4	very strong	27.17–27.88	920–44	131–55	183–217	13–18	4.0–5.5
5	devastating	<27.17	<920	>155	>217	>18	>5.5

Further reading

General and introductory

American Meteorological Society (R. E. Huschke, ed.), *Glossary of Meteorology* (American Meteorological Society, 1959)

B. W. Atkinson and A. Gadd, *A Modern Guide to Forecasting Weather* (Mitchell Beazley, 1986)

R. Chaboud, *How Weather Works* (Thames & Hudson, 1996)

L. Chémery, *Weather and Climates* (Chambers, 2004)

S. Dunlop, *Dictionary of Weather* (Oxford University Press, 2nd edn 2008)

S. Dunlop, *Meteorology Manual* (Haynes, 2014)

D. File, *Weather Watch* (Fourth Estate, 1990)

D. File, *Weather Facts* (Oxford University Press, 1996)

W. Giles, *The Story of Weather* (HMSO, 1990)

J. Gribbin and M. Gribbin, *Watching the Weather* (Constable, 1996)

R. Hamblyn (and Met Office), *The Cloud Book: How to Understand the Skies* (David & Charles, 2008)

R. Hamblyn (and Met Office), *Extraordinary Clouds* (David & Charles, 2009)

R. Hamblyn (and Met Office), *Extraordinary Weather* (David & Charles, 2012)

R. Henson, *Climate Change* (Rough Guides, 2006)

G. Higgins (and Met Office), *Weather World* (David & Charles, 2007); later edn: *Weather Wonders* (David & Charles, 2011)

I. Holford, *The Guinness Book of Weather Facts & Feats* (Guinness Superlatives, 2nd edn 1972)

M. Maslin, *Climate: A Very Short Introduction* (Oxford University Press, 2013)

M. Maslin, *Climate Change: A Very Short Introduction* (Oxford University Press, 2nd edn 2014)

J. Mayes and K. Hughes, *Understanding Weather: A Visual Approach* (Arnold, 2004)

Meteorological Office (R. P. W. Lewis, ed.), *Meteorological Glossary* (HMSO, 6th edn 1991)

Met Office, *The Met Office Book of the British Weather* (David & Charles, 2010)

L. F. Musk, *Weather Systems* (Cambridge University Press, 1988)

C. W. Roberts, *Meteorology*, Yacht Master series (Thomas Reed Publications, 2nd edn 1982)

J. Smith, *The Facts on File Dictionary of Weather and Climate* (Facts on File, 2001)

A. Watts, *The Weather Handbook* (Adlard Coles Nautical, 3rd edn 2014)

J. Williams, *The AMS Weather Book: The Ultimate Guide to America's Weather* (University of Chicago Press, 2009)

L. Young, *Earth's Aura: A Layman's Guide to the Atmosphere* (Penguin, 1997)

Anecdotal and British weather

D. Bowen, *Britain's Weather: Its Workings, Lore and Forecasting* (David & Charles, 1969)

C. Burt, *Extreme Weather* (W. W. Norton, 2007)

A. Dawson, *So Foul and Fair a Day: A History of Scotland's Weather and Climate* (Birlinn, 2009)

S. Dunlop, *Come Rain or Shine* (Summersdale, 2016)

P. Eden, *Great British Weather Disasters* (Continuum, 2008)

P. Eden, *Weatherwise* (Macmillan, 1995)

M. Fish, I. McCaskill, and P. Hudson, *Storm Force* (Great Northern, 2007)

G. Hill, *Hurricane Force* (Collins, 1988)

I. Holford, *British Weather Disasters* (David & Charles, 1976)

J. Kington, *Climate and Weather*, New Naturalist series (Harper Collins, 2010)

D. M. Ludlum, *The American Weather Book* (Houghton Mifflin, 1982)

I. McCaskill and P. Hudson, *Frozen in Time* (Great Northern, 2006)

G. Manley, *Climate and the British Scene*, New Naturalist series (Collins, 1952)

Met Office, *The Climate of England: Some Facts and Figures* (Met Office, 1996)

Met Office, *The Climate of Northern Ireland: Some Facts and Figures* (Met Office, 1996)

Met Office, *The Climate of Scotland: Some Facts and Figures* (Met Office, 1997)

Met Office, *The Climate of Wales: Some Facts and Figures* (Met Office, 1996)

B. Ogley, *In the Wake of the Hurricane: October 16, 1987* (Froglets Publications, 1988)

C. Robinson and E. Findlayson, *Scottish Weather* (Black & White Publishing, 2008)

R. Stirling, *The Weather of Britain* (Giles de la Mare Publishers, 2nd edn 1997)

A. Woodward and R. Penn, *The Wrong Kind of Snow: The Complete Daily Companion to the British Weather* (Hodder & Stoughton, 2007)

Worldwide weather

M. Harding, *Weather to Travel* (Tomorrow's Guides, 3rd edn 2001)

R. Henson, *The Rough Guide to Weather* (Rough Guides, 2nd edn 2007)

E. A. Pearce and C. G. Smith, *The World Weather Guide* (Hodder & Stoughton, 2000)

Historical

W. Akin, *The Forgotten Storm: The Great Tri-State Tornado of 1925* (The Lyons Press, 2002)

N. Courtney, *Gale Force 10* (Review, 2002)

E. Durschmied, *The Weather Factor: How Nature Has Changed History* (Coronet, 2000)

K. Emanuel, *Divine Wind: The History and Science of Hurricanes* (Oxford University Press, 2005)

R. M. Friedman, *Appropriating the Weather: Wilhelm Bjerknes and the Construction of a Modern Meteorology* (Cornell University Press, 1989)

A. Friendly, *Beaufort of the Admiralty: The Life of Sir Francis Beaufort 1774–1857* (Random House, 1977)

J. Gribbin and M. Gribbin, *FitzRoy* (Review, 2003)

P. Halford, *Storm Warning: The Origins of the Weather Forecast* (Sutton, 2004)

R. Hamblyn, *The Invention of Clouds: How an Amateur Meteorologist Forged the Language of the Skies* (Picador, 2001)

S. Huler, *Defining the Wind* (Three Rivers Press, 2004)

R. Inwards, *Weather Lore* (Elliott Stock, 1893 and many later reprints)

P. Jefferson, *And Now the Shipping Forecast* (UIT Cambridge, 2011)

H. Lamb, *Historic Storms of the North Sea, British Isles and Northwest Europe* (Cambridge University Press, 1991)

E. Larson, *Isaac's Storm: The Drowning of Galveston, 8 September 1900* (Fourth Estate, 2000)

M. Monmonier, *Air Apparent: How Meteorologists Learned to Map, Predict and Dramatize the Weather* (University of Chicago Press, 1999)

More specialized works

C. D. Ahrens, *Essentials of Meteorology* (Brooks/Cole, 7th edn 2014)

M. J. Bader et al., *Images in Weather Forecasting* (Cambridge University Press, 1995)

R. J. Barry and R. J. Chorley, *Atmosphere, Weather and Climate* (Routledge, 9th edn 2009)

W. J. Burroughs, *Watching the World's Weather* (Cambridge University Press, 1991)

W. J. Burroughs, *Weather Cycles: Real or Imaginary?* (Cambridge University Press, 2nd edn 2003)

S. Burt, *The Weather Observer's Handbook* (Cambridge University Press, 2012)

R. R. Fotheringham, *The Earth's Atmosphere Viewed from Space* (University of Dundee, 1979)

J. Gleick, *Chaos: Making a New Science* (Vintage, 2nd edn 1997)

E. N. Lorenz, *The Essence of Chaos* (University of Washington Press, 1993)

D. H. McIntosh and A. S. Thom, *Essentials of Meteorology* (Taylor & Francis, 1981)

S. P. Parker (ed.), *Meteorology Source Book* (McGraw-Hill, 1988)

S. H. Schneider (ed.), *Encyclopedia of Climate and Weather*, 2 vols (Oxford University Press, 1996)

R. Simpson (ed.), *Hurricane!* (American Geophysical Union, 2003)

L. Smith, *Chaos: A Very Short Introduction* (Oxford University Press, 2007)

M. Walker, *History of the Meteorological Office* (Cambridge University Press, 2012)

Journals

Weather, Royal Meteorological Society, Reading, UK (monthly)

Weatherwise, Heldref Publications, Washington DC, USA (bimonthly)

Internet links

General information

Atmospheric Optics: <http://www.atoptics.co.uk/>

Hurricane Zone Net: <http://www.hurricanezone.net/>

National Climate Data Centre: <http://www.ncdc.noaa.gov/>

Extremes: <http://www.ncdc.noaa.gov/oa/climate/severeweather/extremes.html>

National Hurricane Center: <http://www.nhc.noaa.gov/>

Reading University (Roger Brugge): <http://www.met.reading.ac.uk/~brugge/index.html>

UK Weather Information: <http://www.weather.org.uk/>

Unisys Hurricane Data: <http://weather.unisys.com/hurricane/atlantic/index.html>

WorldClimate: <http://www.worldclimate.com/>

Meteorological offices, agencies, and organizations

Environment Canada: <http://www.msc-smc.ec.gc.ca/>

European Centre for Medium-Range Weather Forecasting (ECMWF): <http://www.ecmwf.int>

European Meteorological Satellite Organization: <http://www.eumetsat.int/website/home/index.html>

Intergovernmental Panel on Climate Change: <http://www.ipcc.ch>

National Oceanic and Atmospheric Administration (NOAA): <http://www.noaa.gov/>

National Weather Service (NWS): <http://www.nws.noaa.gov/>

UK Meteorological Office: <http://www.metoffice.gov.uk>

World Meteorological Organization: <http://www.wmo.int/pages/index_en.html>

Societies

American Meteorological Society: <https://www2.ametsoc.org/ams/>

Glossary online: <https://www2.ametsoc.org/ams/index.cfm/publications/glossary-of-meteorology/>

Australian Meteorological and Oceanographic Society: <http://www.amos.org.au>

Canadian Meteorological and Oceanographic Society: <http://www.cmos.ca/>

Climatological Observers Link (COL): <https://colweather.ssl-01.com/>

European Meteorological Society: <http://www.emetsoc.org/>

Irish Meteorological Society: <http://www.irishmetsociety.org>

National Weather Association, USA: <http://www.nwas.org/>

New Zealand Meteorological Society: <http://www.metsoc.org.nz/>

Royal Meteorological Society: <http://www.rmets.org>

TORRO: Tornado and Storm Research Organization: <http://torro.org.uk>

Index

Index

155